state of world population 2004

The Cairo Consensus at Ten:
Population, Reproductive
Health and the Global Effort
to End Poverty

United Nations Population Fund
Thoraya Ahmed Obaid, Executive Director

Contents

1 Introduction

On 13 September 1994 in Cairo, after nine days of intense debate, the International Conference on Population and Development (ICPD) adopted a wide-ranging 20-year action plan that delegates and commentators hailed as opening a "new era in population".

Underpinned by a commitment to human rights and gender equality, the Cairo agreement called on countries to ensure reproductive health and rights for all as a critical contribution to sustainable development and the fight against poverty, which the ICPD saw as inseparable from addressing population concerns.

"You have crafted a Programme of Action for the next 20 years, which starts from the reality of the world we live in, and shows us a path to a better reality," Dr. Nafis Sadik, UNFPA Executive Director and Secretary-General of the conference, told delegates at the closing session. "The Programme contains highly specific goals and recommendations in the mutually reinforcing areas of infant and maternal mortality, education, and reproductive health and family planning, but its effect will be far wider-ranging than that. This Programme of Action has the potential to the change the world."

Ten years into the new era, it is time to take stock:

* **The ICPD Programme of Action provides a blueprint for actions** in population and reproductive health that countries agree are **essential to realizing global development goals** including ending extreme poverty and hunger, empowering women, reducing maternal mortality, preserving the environment and stemming the HIV/AIDS pandemic. In recent regional and global meetings and in practice, governments have strongly reaffirmed their commitment, based on experience, to utilize the Programme of Action as an indispensable strategy for improving people's well-being and ensuring human rights.

* **Many developing countries have made great strides in putting the ICPD's recommendations into action**, with a significant impact. Countries are working to integrate population factors with development plans, improve the quality and reach of reproductive health programmes, promote women's rights, meet the needs of young people and those in emergency situations, and strengthen HIV prevention efforts. Access to family planning continues to grow; 60 per cent of married couples in developing countries now use modern methods of contraception, compared to 10-15 per cent in 1960.

* **Inadequate resources and persistent gaps in serving the poorest populations are impeding progress**, however, in meeting ongoing challenges including the continued spread of HIV/AIDS, especially among the young, unmet need for family planning, and high fertility and maternal mortality in the least-developed countries. Donors need to meet the commitments made in Cairo and give due priority to reproductive health in anti-poverty development assistance plans, and programmes must be scaled up and extended to realize the ICPD's goal of comprehensive reproductive health care for all by 2015.

Putting People at the Centre

As its name implied, the ICPD was based on the premise that population size, growth and distribution are closely linked to prospects for economic and social development, and that actions in one area reinforce actions in the other.

This premise had won increasing acceptance in the two decades since the first World Population Conference in 1974, as population grew rapidly in developing regions and as more and more countries gained experience with family planning programmes. By 1994, most developing countries saw a need to address population concerns in order to promote economic growth and improve people's well-being.

A NEW APPROACH. But the Cairo conference radically changed the international community's approach to the interlinked challenges of population and development, putting human beings and human rights, rather than population numbers and growth rates, at the centre of the equation.

At the heart of this paradigm shift was the move away from a perception of population as essentially a macro-economic variable for planning and policy, to a rights-based approach in which the well-being of individuals is key. The ICPD Programme of Action called for policies and programmes to take an integrated approach—linking population action to human development, women's empowerment, gender equality, and the needs and rights of individuals, including young people.

The ICPD Programme of Action recognized that investing in people, in broadening their opportunities and enabling them to realize their potential as human beings, is the key to sustained economic growth and sustainable development, as well as to population levels that are in balance with the environment and available resources.

As part of this shift, the ICPD grounded family planning, once the main focus of population policies and programmes, within a broader framework of **reproductive health and rights, including family planning and sexual health**. It recognized reproductive health as a human right for all people throughout their life cycle, and urged countries to strive for universal access to comprehensive reproductive health services by 2015 (see Chapter 6).

INDIVIDUAL CHOICE PROMOTES PROSPERITY. The ICPD consensus recognized that enabling couples and individuals to freely determine the number, timing and spacing of their children would speed progress towards smaller families and slower population growth, contributing to economic growth and reducing poverty, at both the household and macro levels. Conversely, it understood that not addressing needs and major gaps in reproductive health services would help perpetuate high fertility, high maternal mortality and rapid population growth, undermining poverty reduction prospects (see Chapter 2).

WOMEN'S RIGHTS. Empowering women was recognized as an important end in itself, as well as a key to improving the quality of life of everyone. Without the full and equal participation of women, there can be no sustainable human development. The Programme of Action stressed the importance of reproductive rights to women's autonomy, as a complement to education, economic empowerment and political participation (see Chapter 5).

Important breakthroughs were made in facing up to urgent but sensitive challenges including adolescents' sexual health, HIV/AIDS and unsafe abortion. Unprecedented attention was given to underserved groups, including the rural poor, indigenous peoples, urban slum dwellers, and refugees and internally displaced people.

PARTICIPATION AND PARTNERSHIP. The Cairo agreement also envisioned a participatory and accountable development process, actively involving beneficiaries to ensure that programmes and policy goals are linked

> ## A WIDE MANDATE
>
> *The 1994 Conference was explicitly given a broader mandate on development issues than previous population conferences, reflecting the growing awareness that population, poverty, patterns of production and consumption and the environment are so closely interconnected that none of them can be considered in isolation.*
>
> —ICPD Programme of Action, para. 1.5

with personal realities, and to building broad partnerships between governments, international organizations and civil society.

From Words to Action

As recent country, regional and global reports and surveys make clear, the Cairo agreement has shaped policies and actions over the past ten years addressing a broad range of concerns related to population, reproductive health and gender equality.

At a series of regional conferences marking ten years since the ICPD, governments collectively have strongly reaffirmed their commitment to the Programme of Action, despite efforts by the United States to reopen issues resolved by consensus in Cairo. Many developing countries and countries in transition confirm their national ownership of the ICPD, and report that policies and programmes promoting reproductive health and gender equality are now indispensable parts of their development plans.

And at the Commission on Population and Development's annual session in 2004, countries agreed that implementation of the ICPD Programme of Action makes "an essential contribution to the achievement of internationally agreed development goals, including those contained in the United Nations Millennium Declaration". (See Chapter 2 for more on the **Millennium Development Goals** and their connection to the ICPD.)

Countries Report on Progress

A global survey[1] of governments undertaken in 2003 by UNFPA, the United Nations Population Fund, provides further evidence that developing countries today strongly feel a strong sense of "ownership" of the ICPD agreement, and are making concerted efforts to implement its recommendations and achieve its goals.

Solid gains have been made in integrating population concerns into development strategies to alleviate poverty, promote human rights and redress inequality, protect the environment and conserve natural resources, and decentralize planning. Institutions and laws have been established and modified to speed progress.

Access to reproductive health and family planning services has expanded significantly, along with actions to meet young people's needs, address HIV/AIDS and reduce maternal mortality. But countries responding to the survey also recognized that much more must be done to ensure reproductive rights, access to reproductive health services by adolescents, a wider range of contraceptive choices and higher quality client-centred services.

Among the main findings of UNFPA's global survey:

POPULATION AND DEVELOPMENT. Nearly all (96 per cent) of the 151 developing countries responding reported action to integrate population concerns into development policies and strategies. Most said they had adopted policies to address population-poverty interactions. Half the countries reported activities to influence the distribution of their populations, for example by creating new economic growth centres and decentralizing planning and political decision-making. Countries are also becoming more pragmatic in focusing resources and addressing priority needs (see Chapter 2).

GENDER EQUALITY AND WOMEN'S EMPOWERMENT. Ninety-nine per cent of countries reported that they had adopted policies, laws or constitutional provisions to protect the rights of girls and women. Many have established national commissions for women. Countries have set up mechanisms to provide women with education, skills and employment, and to promote women's equal participation in the political process and community affairs. Laws have been adopted and advocacy undertaken to counter gender-based violence. Various measures have been taken to increase girls' enrolment in primary and secondary schools.

REPRODUCTIVE HEALTH AND FAMILY PLANNING. Countries have begun to integrate reproductive health services into primary health care. Most are improving the training and increasing the numbers of health providers. They have improved service facilities and expanded access, particularly for people living in remote areas. Use of modern contraception continues to grow, and countries have linked family planning with other reproductive health services. Efforts to reduce maternal deaths and injuries are getting increased attention, with more emphasis on attended delivery and expanding the availability of emergency obstetric care and referral and transport systems (see Chapter 6).

HIV/AIDS. Three fourths of countries reported adopting national strategies on HIV/AIDS; a third said they had specific strategies aimed at high-risk groups. Many countries are promoting the consistent and correct use of condoms and providing voluntary counselling and testing. Advocacy campaigns have used celebrities or religious leaders to promote safer sexual behaviour (see Chapter 8).

ADOLESCENTS AND YOUNG PEOPLE. Countries increasingly recognize the need to address the reproductive health and rights of adolescents, and 92 per cent reported action in this regard. Some have rescinded laws and policies that restricted adolescents' access to reproductive health information and services, and more than half have established youth-friendly services. Most have introduced reproductive health education, as an important component of basic life skills, into school curricula and programmes for out-of-school youth. Many countries are also adopting a holistic approach that deals with the larger context of young people's lives, including socio-economic realities, poverty and livelihoods (see Chapter 9).

PARTNERSHIPS. Most governments are working with a wide variety of civil society and private sector groups—including national and international non-governmental organizations (NGOs), particularly family planning associations, women's associations and community groups—on a broad range of ICPD-related issues. This collaboration is especially helpful in reaching groups otherwise not covered by services (see Chapter 11).

National Ownership and Culture

What makes the Cairo consensus work in practice is that each country decides for itself which actions and policies to carry out, based on its own priority needs, cultural imperatives and values. The Programme of Action stresses that implementation of its recommendations "is the sovereign right of each country, consistent with national laws and development priorities, with full respect for the various religious and

| 1 | **CULTURAL SENSITIVITY IN UNFPA PROGRAMMING** |

To be successful and sustainable, development efforts need to recognize local social and cultural realities and promote open dialogue and community involvement. This understanding informs UNFPA's support to countries in implementing the ICPD Programme of Action.

Partnerships with community leaders and institutions are critical to addressing culturally sensitive issues, as the Fund's experience has confirmed.

In Uganda, for instance, female genital cutting among the Sabiny minority was sharply reduced with UNFPA support, by partnering with Sabiny elders to develop alternative rites that reinforced the community's cultural dignity while protecting the human rights of girls.

In Guatemala, which has one of the highest maternal mortality ratios in Latin America, UNFPA helped facilitate an alliance that successfully pushed for the adoption of a groundbreaking law promoting better health for women and their families; it did this by finding common ground among ideologically diverse groups including Catholic leaders, evangelical Christians and the business community.

ethical values and cultural backgrounds of its people, and in conformity with universally recognized international human rights".

UNFPA is the largest multilateral supporter of population and reproductive health programmes in developing countries and the lead UN agency for implementing ICPD recommendations. Like other donors, it provides financial and technical assistance at the request of governments in response to nationally identified priorities.

Birth of a New Global Consensus

AN EVOLVING UNDERSTANDING.[2] The Cairo consensus centred on reproductive health and rights grew out of more than 25 years of experience with population programmes, and evolving international understandings about development and human rights. In 1969, when UNFPA became active, there was no working agreement on population among the members of the United Nations; by 1994 UNFPA had programmes in 140 countries.

At the time of the first World Population Conference in Bucharest in 1974, a large group of countries, including most of Latin America, francophone Africa and parts of Asia, were ambivalent about population activities beyond data gathering and maternal and child health. Two decades later, almost all countries supported the spectrum of reproductive health activities, including voluntary family planning, safe motherhood, HIV/AIDS prevention, and protection against and treatment of sexually transmitted infections.

FAMILY PLANNING AS A HUMAN RIGHT. The International Conference on Human Rights in Tehran in 1968 was the first international forum to agree that "parents have a basic human right to determine freely and responsibly the number and spacing of their children".

The Bucharest conference affirmed that family planning was a right of all "individuals and couples". But its discussion about reducing high rates of fertility in developing countries was not explicitly grounded in women's rights. The 1974 World Population Plan of Action, an uneasy compromise, mentioned women only once.

A year later, however, the First World Conference on Women, in Mexico City, agreed that the right to family planning is essential to gender equality.

The 1984 International Conference on Population, also in Mexico City, added that men should share responsibility for family planning and child-rearing "in order to provide women with the freedom to participate fully in the life of society", an objective "integral to achieving development goals, including those related to population policy".

The 1984 conference also called attention to the large "unmet needs for family planning" among couples who wanted to limit or space child-bearing but lacked access to contraception, and noted that those needs would rise sharply as the number of reproductive-age couples grew in the decade ahead.

The 1992 United Nations Conference on Environment and Development identified rapid population growth as a serious obstacle to sustainable development. But there was no consensus on actions to address it, in part because of lingering distrust of family planning programmes.

This impasse was broken at Cairo by linking development goals to human rights and the advancement of women.

REPRODUCTIVE RIGHTS. During the two decades prior to 1994, a number of international forums had

ICPD ON REPRODUCTIVE RIGHTS

[R]eproductive rights embrace certain human rights that are already recognized in national laws, international human rights documents and other consensus documents. These rights rest on the recognition of the basic right of all couples and individuals to decide freely and responsibly the number, spacing and timing of their children and to have the information and means to do so, and the right to attain the highest standard of sexual and reproductive health. It also includes their right to make decisions concerning reproduction free of discrimination, coercion and violence, as expressed in human rights documents. In the exercise of this right, they should take into account the needs of their living and future children and their responsibilities towards the community. The promotion of the responsible exercise of these rights for all people should be the fundamental basis for government- and community-supported policies and programmes in the area of reproductive health, including family planning.

—from ICPD Programme of Action, para. 7.3

broken new ground in elaborating human rights, including the rights to development and health, women's rights and reproductive decision-making. The ICPD put these together, elaborating a new concept of reproductive rights.

Reproductive rights broadly encompass the right to *reproductive and sexual health*, throughout the life cycle; *reproductive self-determination*, including the rights to voluntary choice in marriage, and to have the information and means to determine the number, timing and spacing of one's children; *equality and equity* for women and men in all spheres of life; and *sexual and reproductive security*, including freedom from sexual violence and coercion.[3] These were spelled out in a variety of human rights treaties and conventions and international consensus agreements.

In the run-up to the ICPD, reproductive rights proponents in governments and civil society, particularly women's groups, mobilized to ensure that these understandings would underpin the new plan of action to address population and development concerns.

The 2003 UNFPA global survey found that since the ICPD, 131 countries had changed national policies or laws, or made institutional changes to recognize reproductive rights. For example, South Africa and Venezuela include reproductive rights in their constitutions as fundamental human rights. India's human rights commission has adopted a declaration on reproductive rights and directed state governments to promote and protect them.[4]

In Ecuador, Ethiopia, Ghana, Kenya and Liberia, among other countries, NGOs such as women's lawyers' organizations promote and monitor government actions supporting reproductive rights.

ABORTION COMPROMISE. Broad agreement was reached on most elements of the Programme of Action in lengthy preparatory meetings. But at the Cairo conference itself, the widely reported disagreement over how to address the abortion issue threatened to block the consensus.

After prolonged debate, a compromise was reached. The 1984 International Conference on Population had agreed that abortion should never be promoted as a means of family planning, that women should be helped to avoid abortion through improved access to family planning, and that those who have had recourse to abortion need humane treatment and counselling. The 1994 Programme of Action reaffirmed these points. Acknowledging that unsafe abortion[5] is a major public health concern, it added that women should have access to quality services for managing complications of abortion. Abortion policy, governments agreed, is a matter for national decision-making; where abortion is not against the law, it should be safe.

The 1995 Fourth World Conference on Women in Beijing upheld the Cairo action plan and reproductive rights as central to the agenda for advancing gender equality. Beijing elaborated on the concept of sexual security, specifying that "The human rights of women include their right to have control over, and decide freely and responsibly on matters related to their sexuality, including sexual and reproductive health, free of coercion, discrimination and violence."

Wide-ranging Impact

The ICPD's success in advancing a human rights agenda to address critical health and development challenges has had a broad impact. Since 1994, NGOs, countries and the international community have used the consensus to help advance a far-reaching agenda on empowerment and equality. For example:

- Countries have stepped up efforts to fight **HIV/AIDS** using an integrated, comprehensive approach to prevention, treatment, care and support (see Chapter 8).

- **Adolescent reproductive health** has become an emerging worldwide concern (see Chapter 9).

- **Early marriage** is increasingly being opposed as a risk to girls' health and a violation of their rights.

- The persistence of high **maternal mortality** has sparked an intensified examination of its causes and remedies (see Chapter 7).

- There is growing recognition of and support for women's **reproductive health needs in emergency situations** (see Chapter 10).

- The UN Security Council in October 2000 unanimously adopted Resolution 1325 on **women, peace and security**, calling for the special needs of women and girls to be incorporated in all decisions related to repatriation and resettlement, rehabilitation, reintegration and post-conflict reconstruction.

- A growing number of countries are taking action to prevent and repair **obstetric fistula**, a terrible injury that happens during childbirth, especially to adolescent mothers.

- Action to reduce the impact of **unsafe abortion**— including greater support for quality family planning and post-abortion care, and open discussion informed by cultural values on the circumstances when abortion might be permissible—has grown since the ICPD identified it as a major public health concern.

- Campaigns against **gender-based violence** (including domestic violence and the use of rape to terrorize war-affected populations) are gaining broad support in many countries.

- Calls are increasing for global action to combat **trafficking in women and children** and coercion in the global sex industry.

- Strong action is being taken to end **female genital cutting** (FGC) and other harmful traditional practices once considered too sensitive to discuss; 17 countries have outlawed FGC and many communities have been mobilized to eliminate the practice.

Long Way to Go

The progress countries have made to date in putting the ICPD's recommendations into practice has laid the groundwork for further advances in ensuring reproductive health and rights. But the challenges remaining to be addressed are daunting:

- **Migration** continues from rural areas of developing countries to fast-growing cities. **By 2007, half the world's population will be urban**. Providing social services including reproductive health care in poor urban areas is a major challenge, as is meeting the needs of underserved rural communities.

- Population growth is contributing, along with high resource consumption by affluent populations, to increasing **stress on the global environment**. Global warming, deforestation, growing scarcity of water and diminishing crop land will make it harder to address poverty and gender inequality.

- **More than 350 million couples still lack access to a full range of family planning services**. Some 137 million women want to delay their next birth or avoid another but are not using family planning; 64 million more are using less-effective methods. Services are reaching many more women than ever before, but are not expanding fast enough to close existing gaps or to keep pace with population growth and rising demand. **Demand for family planning services will increase by 40 per cent by 2025.**

- **Complications of pregnancy and childbirth** are among the leading causes of death and illness among women of reproductive age in many parts of the developing world. Some 8 million women each year suffer life-threatening pregnancy-related complications; over 529,000 die as a consequence, 99 per cent of them in developing countries.[6] Many times that number suffer infection or injury.

- In the developing world as a whole, **one third of all pregnant women receive no health care during pregnancy**; 60 per cent of deliveries take place outside of health facilities; and skilled personnel assist only half of all deliveries.

- There were an estimated **5 million new HIV infections** during 2003, an average of 14,000 per day; 40 per cent were in women and nearly 20 per cent in children. In mid-2004, about 38 million people were living with HIV/AIDS. In 2003, some 3 million people died of AIDS: 2.5 million adults, and 500,000 children under 15.[7]

RESOURCES FALL SHORT. In the face of these challenges, the response of the international community

2 POPULATION GROWTH IS STILL AN ISSUE

Global population, now 6.4 billion, is still growing rapidly—currently by 76 million persons per year. By 2050, the United Nations projects the world will add some 2.5 billion people, an amount equal to the world's total population in 1950.

Growth has slowed since it peaked in the mid-1990s at around 82 million annually. The average family size has declined from six children per woman in 1960 to around three today, as family planning has become more accessible and widely used. Projections suggest total population will start to level off by the middle of this century, as fertility drops to replacement level or lower.

But some countries will reach that point much later than others. Those with young populations (due to high fertility in the recent past) will continue to grow for decades even with small families as the norm. The number of adolescents, aged 10-19, is at an all-time high of 1.2 billion.

And in the poorest countries, where fertility and mortality remain high and access to family planning is limited, the transition to smaller families is only just beginning. The 50 least-developed countries are expected to grow by 228 per cent, to 1.7 billion by 2050.

Countries where fertility has fallen sharply will see a dramatic ageing of their populations in the decades ahead, a trend already well under way in developed countries and a major policy concern.

Ninety-six per cent of the projected growth will be in developing countries. The populations of Europe and Japan are now declining and the pace of decline is projected to double by 2010-2015; North America continues to grow at about 1 per cent annually, mostly because of immigration.

Today's population estimates and growth projections are lower than those made a decade ago, largely because the impact of HIV/AIDS in Africa has been worse than previously projected and growth in the developed countries has declined faster.

The 38 African countries most affected by HIV/AIDS are projected to have 823 million people in 2015, 91 million fewer than if no AIDS deaths had occurred but over 50 per cent more than today (without HIV/AIDS they would have grown by 70 per cent).

The United Nations' projections of slower population growth assume that more couples will be able to choose to have smaller families; this will require greater investments to ensure wider access to reproductive health information and services, including family planning.

Figure 1: World population, 1950-2050 (projected)

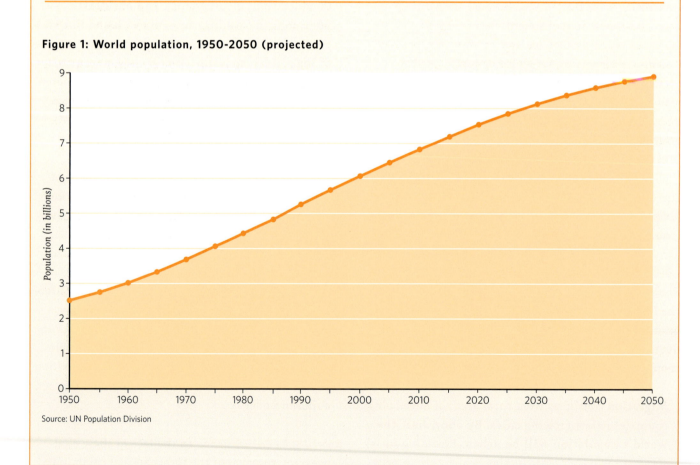

Source: UN Population Division

has been inadequate. After an initial surge following Cairo, resource levels have remained static.

Donor countries have made available only about half of the external resources that the ICPD agreed would be needed to implement the Programme of Action. Donors agreed to provide $6.1 billion a year for population and reproductive health programmes by 2005, a third of the total resources needed. Between 1999 and 2001 their contributions stayed at around $2.6 billion; in 2002 they increased to $3.1 billion.[8]

In the face of the HIV/AIDS pandemic, there are additional needs, particularly for a reliable and sufficient supply of reproductive health commodities, including male and female condoms.

The Way Forward

The tenth anniversary of the ICPD is an opportunity for governments and the international community to review implementation efforts, renew pledges and identify priorities and remaining challenges. Regional reviews and responses to UNFPA's global survey have confirmed that countries have made significant progress, and are strongly committed to further action.

With its comprehensive approach linking population and development—including environmental protection and the management of urban and rural growth—gender equality, and reproductive health and rights, the Programme of Action continues to offer an essential blueprint for development efforts in the coming decade.

Recent commitments by the United Nations and donors to poverty reduction strategies and the Millennium Development Goals (including action to reduce maternal mortality and stem the HIV/AIDS pandemic) offer a real chance to generate the additional political will and resources that will be needed to fully implement the Cairo consensus.

The ICPD goal of universal access to reproductive health care by 2015 is an essential condition for meeting most of the MDGs. It is critical to ensure that resources and actions needed for reproductive health

are not overlooked when funding priorities are set. Donor support in this sector is only about half the level that the ICPD agreed on, and needs continue to increase.

Additionally, funding is needed for integrated, multisectoral programmes. This approach, at the heart of the vision of the ICPD, contrasts with the sector-by-sector (and within health, disease-by-disease) programming approaches that the orientation of the MDGs has facilitated.

Investments in better reproductive health have a proven high return. More funding is needed, in particular, to increase the availability of voluntary family planning services, to expand access to emergency obstetric care and other safe motherhood interventions, and to dramatically scale up HIV/AIDS prevention efforts as part of an intensified response to the pandemic. Special efforts are needed to reach adolescents and young people, and those displaced by wars and natural disasters.

It is also important to reinforce other fundamental ICPD conclusions: development plans and policies need to address population dynamics and its link with reproductive health, and their impact on prospects for sustainable economic growth and poverty reduction; communities and beneficiaries must be involved in shaping and evaluating programmes; and interventions must be carried out in partnership with civil society and be culturally sensitive.

Ten years ago, the ICPD Programme of Action began by noting that the world was "at a defining moment in the history of international cooperation", an unparalleled chance to advance human well-being by linking development to population, women's advancement and reproductive health. Today's challenges—including security concerns, the continuing spread of HIV/AIDS, and persistent poverty alongside unprecedented prosperity—make it all the more imperative to carry out the Cairo agenda so its dream of a better future can be realized.

2 Population and Poverty

A central premise of the 1994 Cairo conference was the notion that the size, growth, age structure and rural-urban distribution of a country's population have a critical impact on its development prospects, and specifically on prospects for raising the living standards of the poor. Reflecting this understanding, the ICPD called on countries to "fully integrate population concerns into development strategies, planning, decision-making and resource allocation at all levels".

Among the key population-development concerns the Programme of Action addressed were: population and poverty; the environment (see Chapter 3); health, morbidity and mortality (Chapters 6, 7 and 8); and population distribution, urbanization and internal and international migration (Chapter 4).

Poverty perpetuates poor health, gender inequality and rapid population growth. The ICPD recognized that empowering individual women and men with education, equal opportunity and the means to determine the number and spacing of their children is critical to breaking this vicious cycle.

In 1994 there was already solid evidence, based on two generations of experience, that developing countries with lower fertility and slower population growth have higher productivity, more savings and more productive investment, resulting in faster economic growth.

Analysis of more recent data confirms that countries that have reduced fertility and mortality by investing in health and education have prospered as a result.

As the international community strives to focus development efforts more effectively to achieve the **Millennium Development Goals** for eradicating poverty and improving people's well-being, the ICPD's rights-based agenda for addressing the interdependence of population and poverty deserves the highest priority.

Millennium Development Goals

In the decade since the ICPD, policies shaping international development assistance have changed. The amount of assistance has stagnated at around $60 billion per year, a result of both donor fatigue and economic uncertainty. At the same time, donors have become more critical of how assistance has been used (with blame falling on both donor and recipient governments).

To increase the impact of development assistance, donors have made governance an important criterion for its allocation, and strengthened the overall focus on alleviating poverty as the main rationale for assistance.

The aim of focusing development assistance more effectively shaped the Millennium Summit at UN Headquarters in 2000 and its identification of the Millennium Development Goals (MDGs) and associated targets for reducing global poverty by 2015:

1. **Eradicate extreme poverty and hunger**. By 2015, halve the proportion of people living on less than a dollar a day and those who suffer from hunger.

2. **Achieve universal primary education**. By 2015, ensure that all boys and girls complete primary school.

3. **Promote gender equality and empower women**. Eliminate gender disparities in primary and secondary education preferably by 2005 and at all levels by 2015.

4. **Reduce child mortality**. By 2015, reduce by two thirds the mortality rate among children under 5.

5. **Improve maternal health**. By 2015, reduce by three quarters the ratio of women dying in childbirth.

6. **Combat HIV/AIDS, malaria and other diseases**. By 2015, halt and begin to reverse the spread of HIV/AIDS and the incidence of malaria and other major diseases.

7. **Ensure environmental sustainability**. Integrate the principles of sustainable development into country policies and programmes and reverse the loss of environmental resources. By 2015, reduce by half the proportion of people without access to safe drinking water. By 2020, achieve significant improvement in the lives of at least 100 million slum dwellers.

8. **Develop a global partnership for development**. Develop further an open trading and financial system that includes a commitment to good governance, development and poverty reduction—nationally and internationally. Address the least-developed countries' special needs, and the special needs of landlocked and small island developing states. Deal comprehensively with developing countries' debt problems. Develop decent and productive work for youth. In cooperation with pharmaceutical companies, provide access to affordable essential drugs in developing countries. In cooperation with the private sector, make available the benefits of new technologies—especially information and communications technologies.

In many ways, the goals and targets set at the ICPD (see Box 3) anticipated the MDGs.

Reproductive Health and the MDGs

The Cairo goal of universal access to quality reproductive health services by 2015 is not one of the MDGs. This has led to concern that reproductive health might get short-changed in efforts to better direct resources to development priorities. But as the ICPD affirmed, this goal is fundamental to reducing poverty, child and maternal mortality, and the spread of HIV/AIDS.

As UN Secretary-General Kofi Annan stated in a message to the Fifth Asian and Pacific Population Conference, held in Bangkok in December 2002,

---

3 GOALS OF THE ICPD AND THE 1999 REVIEW

The ICPD adopted the following mutually supportive goals:

- **Gender equality in education.** Eliminate the gender gap in primary and secondary education by 2005, and complete access to primary school or the equivalent by girls and boys as quickly as possible and in any case before 2015;

- **Infant, child and maternal mortality.** Reduce infant and under-5 mortality rates by at least one third, to no more than 50 and 70 per 1,000 live births, respectively, by 2000, and to below 35 and 45, respectively, by 2015; reduce maternal mortality to half the 1990 levels by 2000 and by a further one half by 2015 (specifically, in countries with the highest mortality, to below 60 per 100,000 live births);

- **Reproductive health services.** Provide universal access to a full range of safe and reliable family-planning methods and to related reproductive and sexual health services by 2015.

Reviewing the first five years of implementing the Programme of Action, the United Nations in 1999 took note of the worsening crisis of HIV/AIDS and the vulnerability of young people and adopted specific numerical targets to evaluate programme implementation:

- **Education.** Halve the 1990 illiteracy rate for women and girls by 2005; ensure that by 2010 at least 90 per cent of children of both sexes are enrolled in primary school;

- **Reproductive health services.** Provide a wide range of family planning methods, essential obstetric care, and prevention and management of reproductive tract infections in 60 per cent of primary health care facilities by 2005; in 80 per cent by 2010, and in all by 2015;

- **Maternal mortality.** Where maternal mortality is very high, ensure that at least 40 per cent of all births are assisted by skilled attendants by 2005, 50 per cent by 2010 and 60 per cent by 2015; globally, 80 per cent of births should be attended by 2005, 85 per cent by 2010 and 90 per cent by 2015;

- **Unmet need for family planning.** Reduce by half by 2005 any gap between the proportions of individuals using contraceptives and those expressing a desire to space or limit their families, by 75 per cent by 2010, and completely by 2015. Recruitment targets or quotas should not be used to reach this goal.

- **HIV/AIDS.** Ensure that by 2005 at least 90 per cent, and by 2010 95 per cent, of young men and women 15-24 have access to HIV/AIDS prevention methods such as female and male condoms, and voluntary testing, counselling and follow-up; reduce HIV infection rates in this age group by 25 per cent in the most-affected countries by 2005, and by 25 per cent globally by 2010.

"The Millennium Development Goals, particularly the eradication of extreme poverty and hunger, cannot be achieved if questions of population and reproductive health are not squarely addressed. And that means stronger efforts to promote women's rights, and greater investment in education and health, including reproductive health and family planning."

Much more needs to be done to ensure synergy between the MDGs and the goals of the ICPD, but encouraging progress has been made. Two of the UN Millennium Project's expert task forces (in the areas of gender equality and child and maternal health) have strongly endorsed "universal access to sexual and reproductive health" as a strategic priority for attaining the MDGs.[1]

All countries are required to report to the General Assembly on progress towards the MDGs, through National Millennium Development Goals Reports. Ten of the first reports published listed reproductive health as a goal, and an additional four wrote about reproductive health issues. Nine provided data on the contraceptive prevalence rate (the most frequently used indicator to monitor access to reproductive health care); 10 others made reference to it.

Economic Impact of Population Dynamics

There is clear evidence that enabling people to have fewer children, if they want to, helps to stimulate development and reduce poverty, both in individual households and at the macro-economic level.

FAMILY SIZE AND WELL-BEING. Recent research supports the premise that having many (and unplanned for) children imposes a heavy burden on the poor, while smaller families have higher upward economic mobility.[2]

Fertility impacts on a family's poverty in several ways:

- Smaller families share income among fewer people, and average income per capita increases. A family of a certain size may be below the poverty line, but with one less member may rise above the poverty threshold.

- Fewer pregnancies lead to lower maternal mortality and morbidity, and often to more education and economic opportunities for women. A mother's death or disability can drive a family into poverty. Her ability to earn income can lead the family out of poverty.

- High fertility undermines the education of children, especially girls. Larger families have less to invest in the education of each child. In addition, early pregnancy interrupts young women's schooling, and in large families mothers often remove daughters from school to help care for siblings. Less education typically implies increased poverty for the family as well as the inter-generational transmission of poverty.

- Families with lower fertility are better able to invest in the health of each child, and to give their children proper nourishment. Malnourishment leads to stunted growth, cerebral underdevelopment and subsequent inability to achieve high levels of productivity in the labour force.

MACRO-ECONOMIC IMPACT. High fertility impedes development in a variety of ways. The World Health Organization (WHO) Commission on Macroeconomics and Health noted in 2001, "At the societal level, rapid rural population growth in particular puts enormous stress on the physical environment and on food productivity as land-labour ratios in agriculture decline. Desperately poor peasants are then likely to crowd cities, leading to very high rates of urbanization, with additional adverse consequence in congestion and in declining urban capital per person."

Lower fertility, on the other hand, is linked to economic gains. A 2001 study of 45 countries found that if these countries had reduced fertility by 5 births per 1,000 people in the 1980s, the average national incidence of poverty of 18.9 per cent in the mid-1980s would have been reduced to 12.6 per cent between 1990 and 1995.[3]

At the time of Cairo, econometric proof of this "population effect" on economic growth was difficult to obtain, and mainstream economists tended to dismiss it or play down its importance. A 1986 study by the National Research Council in the United States[4] concluded that population growth had little or no effect on overall economic growth, despite its impor-

INEQUALITY IN REPRODUCTIVE HEALTH FOSTERS POVERTY

The ICPD recognized that ill health and unplanned births can determine whether a family falls into or escapes poverty, as the poor themselves have long known. But policy makers have been slow to address the inequitable distribution of health information and services that helps keep people poor.

An analysis of data on access to reproductive health among different income groups in 56 countries shows that the poorest groups are clearly disadvantaged, in a number of ways:

• The biggest gap between richer and poorer populations is in delivery by a skilled attendant, the most expensive of the reproductive health services;

• Adolescent fertility showed the next largest differential—poorer women have children at younger ages;

• Wealth-based health inequities are greater for safe motherhood, adolescent fertility, contraceptive use and total fertility than for infant mortality;

• Poor women have more children throughout their lives than wealthier women;

• Poor countries have a heightened risk of maternal, infant and child death and illness, and poor women in all countries face higher risks than others;

• Use of family planning, particularly of modern methods, is higher in richer segments of society.

These findings corroborate those presented in *The State of World Population 2002*, which examined data from 44 countries. Shortages of resources, skills, opportunities and outreach deprive the poor of access to reproductive health information and services and the effects are apparent.

The information and service deficits result from various factors:

• Poor women and couples have less access to information and to the skills education provides to expand their knowledge;

• Poor individuals and communities are risk-averse—less likely to try new behaviours—since their room for error is so small;

• Costs for information and services (formal and informal monetary costs, and transport and opportunity costs) are more daunting for the poor;

• When addressing the poor, service providers are less willing or able to interact as closely as is required to exchange information and support about sensitive topics;

• Services are not in locations or open at times accessible to the poor;

• Richer populations are more skilled at working with formal institutions and receiving a responsive hearing.

In 2000, only 3 per cent of gross domestic product was devoted to the health sector in developing countries; in the least-developed countries the figure was even lower. Expenditures in many countries still tend to favour hospitals and medical facilities in the capital city, and there has been little progress towards a more equitable distribution of resources at local levels: the percentage of national health expenditures devoted to local health services has stagnated in developing countries and decreased in the least developed.

---

tant effects at the household level; but it relied on data from the 1960s and 1970s, when many developing countries were early in their demographic transition.

THE DEMOGRAPHIC WINDOW. A new round of research in the mid-1990s,[5] using data from longer periods, showed clearly that falling fertility opens a "demographic window" of economic opportunity. With fewer dependent children relative to the working-age population, countries can make additional investments which can spur economic growth and help reduce poverty.

This window opens only once and closes as populations age and the ratio of dependants (children and the elderly) eventually starts to rise again.

Several countries in East Asia—the so-called Asian Tigers—and a few others have taken advantage of this economic bonus. China has seen a dramatic drop in the incidence of poverty.[6] One study estimated that

declining fertility in Brazil has raised the annual growth of GDP per capita by 0.7 percentage points. Mexico and other Latin American countries have registered similar effects. On the other hand, some countries have largely squandered the opportunity for a one-time "windfall" because of a lack of good governance or policies that have led to unproductive investments.[7]

In the poorest countries where fertility remains high, the demographic window will not open for some time, but investments now—particularly in improving reproductive health service delivery—could hasten its arrival and ensure future dividends.

The world's regions are at different stages of the demographic transition. **South Asia** will reach its peak ratio of working-age to dependent-ages between 2015 and 2025. In **Latin America and the Caribbean**, the proportion in working ages started to increase earlier than in East Asia and will peak during 2020-

2030, but the proportional change has been less marked, and the economic bonus will be correspondingly less sudden and less intense. Some **Arab** and **Central Asian** countries will approach their demographic opportunity within two decades, while others are farther away.

In much of **sub-Saharan Africa** the demographic bonus is still a long way off. The population is still very young and the proportion in working ages relatively low. Many countries are just beginning the demographic transition, and others have not even started. Only 11 countries are projected to reach their maximum working-age proportion before 2050. Unmet need for contraception in the region is high, however, suggesting the expansion of quality programmes could hasten the arrival of the bonus.

HIV/AIDS and Poverty

In countries with high HIV/AIDS prevalence, the pandemic is killing large numbers of people in their most productive years, increasing the ratio of dependents to working-age populations and creating a worst-case scenario with respect to the demographic transition.

For the seven African countries with adult HIV prevalence of 20 per cent or more, the population is projected to be 35 per cent lower by 2025 than it would have been in the absence of AIDS. By 2020-2025, life expectancy in these countries will be 29 years less than it would be without AIDS, a 41 per cent difference.[8]

There is also growing evidence that per capita economic growth will be diminished as a result of increasing dependency ratios, increased burdens on health systems, constrained investment in productivity and reduced labour forces.

Ageing Populations

Ageing is another aspect of population dynamics that affects the level of poverty. As fertility declines and the population ages, traditional family-based systems

for providing old-age care may weaken, leaving the elderly vulnerable. The income security of older persons is a policy concern not only in developed countries but increasingly as well in developing countries such as China that have passed through the demographic transition.[9] Between 2000 and 2050, the proportion of the population aged 65 and older will have doubled in most developing regions (see chart).

National Action to Reduce Poverty

Of the 151 developing countries that responded to the UNFPA global survey[10] in 2003, 136 indicated that they had taken into account population-poverty interactions to some degree, although only 77 countries, or 57 per cent, had taken two or more measures to address them. At the time of the last global survey in 1994, only 13 per cent of developing countries reported any action on this issue.

Countries have adopted diverse strategies to address population and poverty, including the adoption of broad population and development measures in 108 countries (79 per cent); the establishment of special strategies for migrants, refugees, internally displaced persons and other vulnerable groups in 39 countries (29 per cent); and the instituting of measures for income generation and women's empowerment in 25 countries (18 per cent). About 20 countries (14 per cent), had strategies to lower fertility levels while a similar number had strategies to reduce population growth.

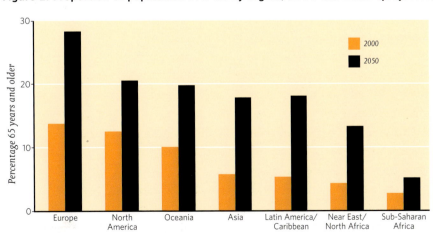

Figure 2: Proportion of population over 65 by region, 2000 and 2050 (projected)

Percentage 65 years and older

2000
2050

Europe | North America | Oceania | Asia | Latin America/ Caribbean | Near East/ North Africa | Sub-Saharan Africa

Source: U.S. Bureau of the Census (2000).

3 Population and the Environment

Stress on the environment and the depletion of natural resources both reinforce and are exacerbated by gender inequality, poor health and poverty, the Cairo conference emphasized. Environmental stress is increasing, due to both "unsustainable consumption and production patterns" (including high resource consumption in wealthy countries and among better-off groups in all countries) and demographic factors such as rapid population growth, population distribution and migration.

Affirming that "meeting the basic human needs of growing populations is dependent on a healthy environment", Chapter III of the ICPD Programme of Action' addressed the interrelationships among population, economic growth and protection of the environment, reiterating principles of Agenda 21, adopted by the United Nations Conference on the Environment and Development in Rio in 1992.

At both the Cairo conference and its five-year review, the global community affirmed that greater equality between men and women is an essential component of sustainable development, including environmental protection. Boosting the status of women is now accepted as a prerequisite for lowering fertility and ensuring sound management of natural resources. And awareness is increasing of the need to address environmental crises, demographic realities, gender inequity and rising consumption amid persistent poverty in a holistic manner.

This understanding has led to a variety of actions since 1994 that link anti-poverty efforts to women's empowerment, health and better management of local resources. Most, however, have been undertaken on a small scale. There is a pressing need to better integrate population, reproductive health and gender-related interventions into strategies for achieving the Millennium Development Goals including MDG 7, ensure environmental sustainability.

The 2003 UNFPA global survey found that countries have made progress in addressing population issues within the context of poverty, environment, and decentralized planning processes. One hundred and twenty-two countries reported developing plans or strategies on population-environment linkages. Forty countries have developed specific policies, and 22 have put in place laws or legislation on population dynamics and the environment.

A VIRTUOUS CIRCLE

Efforts to slow down population growth, to reduce poverty, to achieve economic progress, to improve environmental protection, and to reduce unsustainable consumption and production patterns are mutually reinforcing. Slower population growth has in many countries increased those countries' ability to attack poverty, protect and repair the environment, and build the base for future sustainable development.

—ICPD Programme of Action, para. 3.14

Over the past century and especially over the past 40 years, people have effected vast changes in the global environment. Those most directly affected by environmental challenges, from water pollution to climate change, are also the poorest—and least able to change livelihoods or lifestyles to cope with, or combat, ecological decline. Some snapshots:

- Farmers, ranchers, loggers, and developers have cleared about half the world's original **forest cover**, and another 30 per cent is degraded or fragmented.

- Over the last half century, land degradation has reduced cropland by an estimated 13 per cent and pasture by 4 per cent. In many countries, population growth has raced ahead of food production in recent years. Some 800 million people are chronically malnourished and 2 billion lack food security.

- Three quarters of the world's **fish stocks** are now fished at or beyond sustainable limits. Industrial fleets have fished out at least 90 per cent of large ocean predators—including tuna, marlin and swordfish—in the last 50 years.

- Since the 1950s, global demand for **water** has tripled. Groundwater quantity and quality are declining due to over-pumping, runoff from fertilizers and pesticides, and leaking of industrial waste. Half a billion people live in countries defined as water-stressed or water-scarce; by 2025, that figure is expected to surge to between 2.4 billion and 3.4 billion.

- **Climate change**. As a result of fossil fuel consumption, carbon dioxide levels today are 18 per cent higher than in 1960 and an estimated 31 per cent higher than at the onset of the Industrial Revolution in 1750. Accumulation of greenhouse gases in the atmosphere, including carbon dioxide, is tied to rising and extreme change in temperatures, and more severe storms.

- **Sea level** has risen an estimated 10-20 centimetres, largely as a result of melting ice masses and the expansion of oceans linked to regional and global warming. Small island nations and low-lying cities and farming areas face severe flooding or inundation.

Still, the stakes are high, as human activity continues to alter the planet on an unprecedented scale. More people are using more resources with more intensity and leaving a bigger "footprint" on the earth than ever before.

Population's Impact on Resource Use

Numbers alone do not capture the impact of the interactions between human populations and the environment. The size and weight of the **environmental footprint** each person plants on Earth is determined by the ways people use resources, which affects the quantities they consume. For instance, a vegetarian who primarily uses a bicycle has a much smaller impact than someone who eats meat and drives a sport utility vehicle.

The ecological footprint of an average person in a high-income country is about six times bigger than that of someone in a low-income country, and many more times bigger than in the least-developed countries. The combined footprints of people in a region determine the prospects for saving or permanently losing the biological diversity found there.

Many economists and environmentalists use an equation that ties together population, consumption and technology to describe their relative impacts (I=PAT: Impact=Population x Affluence x Technology).

As birth rates fall, consumption levels and patterns (affluence), coupled with technology, will take on new importance in determining the state of the global environment. But population will remain the critical factor where lack of access to reproductive health services and family planning, shortfalls in education for girls and women, poverty and women's limited power relative to men continue to fuel high fertility.

GLOBAL CONSUMERS AND PERSISTENT POVERTY.

A rapidly growing global consumer class, now around 1.7 billion people, accounts for the vast majority of meat eating, paper use, car driving, and energy use on the planet, as well as the resulting impact of these activities on its natural resources. This class is not limited to industrialized countries; as populations surge in developing countries and as the world economy becomes increasingly globalized, more and more people have the means to acquire a greater diversity of products and services than ever before.[2]

Meanwhile, 2.8 billion people—two in five—still struggle to survive on less than $2 a day. In 2000, 1.1 billion people did not have reasonable access to safe drinking water, and 2.4 billion people worldwide lived without basic sanitation. Lack of access to clean water and sanitation in the developing world led to 1.7 million deaths in 2000.[3]

DIFFERENTIAL IMPACTS. Where population growth and high levels of consumption coincide, as they do in some industrial nations, the impact of growth is significant. For instance, even though the United States' population is only a fourth as large as India's, its environmental footprint is over three times bigger—it releases 15.7 million tons of carbon into the atmosphere each year compared with India's 4.9 million tons.[4] Hence the impact of the current 3 million annual population increase in the United States is greater than that of India's 16 million increase.

Environmental impact can continue to grow even as population growth levels off. In China, population growth has slowed dramatically, but consumption of oil and coal and the resulting pollution continues to rise. While the Chinese Government is promoting greater fuel efficiency for cars (see Box 7), it is not promoting increased use of public transportation, biking and walking, or efficient urban planning so people would not have to drive.

Besides reducing overall resource use, governments can reduce the environmental impacts of increased

consumption by promoting appropriate technology that uses resources more efficiently.[5] Industrial countries can help the developing world by assisting with the dissemination and adoption of cleaner technologies.[6]

Other demographic trends intersect with consumption in surprising ways. As a result of rising incomes, urbanization, and smaller families, the average number of people living under one roof declined between 1970 and 2000—from 5.1 to 4.4 in developing countries and from 3.2 to 2.5 in industrial countries—while the total number of households increased. Each new house requires land and materials. And with fewer people in each household, savings from shared use of energy and appliances are lost. A one-person household in the United States, for example, uses 17 per cent more energy per capita than a two-person household.

Even in some European nations and in Japan, where population growth has stopped, changing household dynamics are important drivers of increased consumption.[7]

Poverty and Ecological Stress

While consumers, particularly in the wealthiest countries, are doing the most to reshape the natural world through their use of resources and products, fast-growing populations in the poorest, least-developed countries also have an impact. Here, biodiversity is often high and environmental degradation already widespread.

Poor populations in many biodiversity-rich regions—largely rural areas where good health facilities, schools, and basic infrastructure are frequently absent—often have no other options but to exploit their local environment to meet subsistence needs for food and fuel.

In these settings, traditional practices that may have been ecologically viable when the population was small are becoming increasingly less viable for species and ecosystems as population grows and demands rise. The trade in bush meat in Central Africa, for instance, has accelerated to such a degree that the future of forest-dwelling animals, including primates, is in jeopardy.[8]

ENVIRONMENT AND HEALTH. From polluted air to contaminated water to toxins in food, the health of

---

| 6 | **PROMOTING MORE EFFICIENT USE OF ENERGY** |

A number of initiatives suggest countries are taking seriously the challenge of reducing harmful production and consumption patterns. For example, China last year began to regulate its rapidly growing auto industry, requiring new family vehicles sold in major cities to meet air pollution standards as strict as those in the United States and Western Europe. Starting this year, new fuel economy standards for cars will be significantly more stringent than those in the United States.

The transfer of energy-efficient technology is also growing. China has become the world's largest manufacturer of efficient compact fluorescent light bulbs, in part through joint ventures with lighting firms in Japan, the Netherlands and elsewhere. India has become a major manufacturer of advance wind turbines using technology obtained through joint ventures and licensing agreements with Danish, Dutch, and German firms.

the environment can affect human health in complex ways. Both women and men are exposed to an increasing number of environmental hazards, especially in poor communities. In rural areas, farmers and labourers often come in contact with an array of pesticides, solvents, and unknown toxins; some of these have an effect on reproductive health, with a disproportionate impact on women.[9]

Gender Dimensions of Environmental Change

In the developing world in particular, gender plays a strong role in how resources are used and developed. Women and girls often spend hours each week fetching water for domestic use, for example; when water supplies are erratic, it is they who suffer the greatest consequences.[10] In Sudan, where deforestation has quadrupled the amount of time women spend gathering wood for cooking, the energy used to tote water and fuel accounts for one third of a woman's daily calorie intake, according to the World Health Organization.[11]

Rights to natural resources are often heavily biased. Few women own property (in some countries they are legally prohibited from doing so) and few are involved in high-level decision-making on the environment. For the most part, men are still largely responsible for deciding how the world's natural resources are used through industry, mining, livestock grazing and land tenure.

Development agencies still offer technical assistance mainly to men, even in places where women are the ones toting the wood and water and tilling the soil. When government officials or community leaders fail to recognize the different ways women use resources—growing vegetables for family consumption in the spaces between male-managed cash crops, for example—the resources are easily destroyed.[12]

But when women are included in natural resource management, the results can be dramatic.

When a water project that excluded women in the Kirinyaga district of Kenya failed, local women formed the Kugeria Women's Group and asked the Ministry of Water Development to help them gain access to safe, affordable sources of water. Their efforts have brought water to 300 families, improved

Before the ICPD, many policy makers tended to view "development" in the restricted sense of economic growth, measured by gross national product. Prescriptions for development were often confined to an economic agenda involving investment, trade negotiations, infrastructure construction and monetary aid. Considerations such as gender equality and equity, health, education and the state of the environment were treated as secondary if addressed at all.

Ten years after the Cairo conference, there is much greater recognition that good stewardship of the environment, people's health and the status of women are all interconnected and bear on the speed and breadth of a country's development. True development must improve the lives of individuals.

Some demographers and scholars concerned with population-development relationships and the environment contend the Cairo conference over-emphasized sexual and reproductive health services and played down the macro-level relationships between population growth and the environment, the economy, poverty reduction, education and housing.

Such criticism is unwarranted. Cairo recognized that promoting individual rights with regard to sexual and reproductive health would lead to macro-level progress as well—that meeting expressed desires and ensuring people's right to chose the number, timing and spacing of their children would slow rapid population growth, without resorting to demographic targets. Indeed, enabling health systems to meet individuals' needs and wishes in a more client-friendly manner could even accelerate family planning use.

---

sanitation, and increased agricultural production. The women have also become community leaders, working to build a clinic and provide access to reproductive health and family planning services.[13]

DEVELOPING INTEGRATED APPROACHES. Following Agenda 21 and the ICPD, there has been greater international attention to women's stewardship of natural resources, including efforts to integrate reproductive health and family planning into conservation programmes. Some environmental groups have developed partnerships with population organizations. For example, Conservation International has teamed up with family planning NGOs and the Mexican Social Security Institute to expand access to reproductive health care including family planning, and to halt the clearing of forests in and around the Montes Azules Biosphere Reserve.

In the mountainous provinces of central Ecuador, where women do not have access to reproductive health services and soil erosion is widespread, World Neighbors has joined with a local NGO, the Centre for Medical Guidance and Family Planning, to deliver reproductive health care and to promote improvements in local management of natural resources to over 4,000 families.

In March 2002 in Helsinki, women environment ministers and representatives from 19 industrial and developing countries met with women's NGOs and issued a statement calling for: equal rights for women in access to and control of natural resources, including land tenure; policies that involve women in decisions about resource use; better consumer education on the environmental impacts of products; and the development of "policies, legislation and strategies towards gender balance in environmental protection and in the distribution of its benefits".[14]

POLICY CHANGES. At the policy level, many countries, drawing on recommendations of the ICPD, its fifth-year review, the Millennium Summit and the 2002 World Summit for Sustainable Development, have emphasized the linkages among population dynamics, sustainable development and environmental protection.

In Azerbaijan, for example, the State Programme on Poverty Reduction and Economic Development takes into account population and environment interrelationships; promotes public education on environmental issues that directly affect population groups; works to monitor environmental impacts of policies at the local and community levels; and emphasizes the protection and preservation of the environment as both a source and an outcome of sustained economic growth.

In the Seychelles, two comprehensive environment management plans have been developed over the past decade that integrate population and development. The latest plan, covering 2000-2010, focuses on urbanization, water management, population and health, gender, environmental economics and sustainable financing.

4 Migration and Urbanization

During the past ten years, migration has increased, both within and between countries, and the phenomenon has grown in political importance.

Recognizing that orderly migration can have positive consequences on both sending and receiving countries, the ICPD Programme of Action (Chapters IX and X) called for a comprehensive approach to managing migration. It emphasized both the rights and well-being of migrants and the need for international support to assist affected countries and promote more interstate cooperation around the issue.

Urbanization and Relocation

By 2007, for the first time in human history, more than half the people in the world will be living in cities, the result of a continuing movement of people that has led to a tremendous growth of urban areas in developing countries in the past decade. Helping countries respond to this population shift was a key priority for the ICPD.

The Programme of Action devoted an entire chapter to the spatial distribution of the population and internal population movements. It recognized that people move within countries in response to the inequitable distribution of resources, services and opportunities. Push factors—particularly rural poverty—and pull factors—the attraction of more economically dynamic urban areas and new land tenure prospects in rural frontiers—contribute to these population movements.

As can be the case for international migration, a significant proportion of internal migration is temporary, for example, with labour migrants returning to their farms during busy seasons.

Like earlier population conferences, the ICPD sought to promote integrated and sustainable development policies to address imbalances within countries and between population growth and economic growth. Action recommendations aimed to improve infrastructure and services for poor, indigenous groups and other underserved rural populations.

Another focus was managing population growth and developing infrastructure in large urban areas. These are urgent challenges for development and for

---

MANAGING MIGRATION

In order to achieve a balanced spatial distribution of production employment and population, countries should adopt sustainable regional development strategies and strategies for the encouragement of urban consolidation, the growth of small or medium-sized urban centres and the sustainable development of rural areas, including the adoption of labour-intensive projects, training for non-farming jobs for youth and effective transport and communication systems. To create an enabling context for local development, including the provision of services, governments should consider decentralizing their administrative systems.

—from ICPD Programme of Action, para. 9.4

improving the lives of the poor, many of whom live in slums and peri-urban settlements with limited access to health care and other services.[1]

The ICPD recognized the economic dynamism of large urban settlements, but also acknowledged the growing importance of medium-sized cities and of migration between cities.[2]

Today, more policy attention is being given to the economic diversity within cities and neighbourhoods, where rich and poor often live in close proximity.[3]

Millennium Development Goal 7, Ensure environmental sustainability, has as a target, "By 2020, achieve significant improvement in the lives of at least 100 million slum dwellers."

The latest estimates and projections indicate a majority of the global population will be urban by 2007.[4] The number of urban dwellers will rise from 3 billion in 2003 (48 per cent of the total population) to 5 billion in 2030 (60 per cent). Most of this urban growth will be due to natural fertility rather than migration. The rural population will decline slightly in the same period, from 3.3 to 3.2 billion.

The urban population is projected to grow by 1.8 per cent per year between 2000 and 2030, almost twice as fast as global population growth. Less-developed regions will grow by 2.3 per cent and are expected to be majority urban by 2017. By 2030 all regions of the world will have urban majorities (Africa will reach 54 per cent urban; Asia, 55 per cent). Almost all of the world's total population growth in this period will be in urban areas of developing countries.

HIV/AIDS has added a new element of uncertainty to these projections.[5] Overall, infection rates have tended to be higher in urban areas. In heavily affected areas, higher urban death rates and lower fertility rates might slow the pace of urbanization or even result in a decline in urban population.

Today there are 20 cities of more than 10 million people (15 in developing countries), containing 4 per cent of the global population; by 2015 there will be 22 such mega-cities (16 in developing countries), with 5 per cent of the global population.

Cities with fewer than 1 million persons will add 400 million people by 2015, and more than 90 per cent of this growth will be in cities of fewer than 500,000.

This will require vast improvements in local infrastructure and in the capacity to manage public services, particularly as decision-making is increasingly being decentralized to local municipalities and districts.

Greater attention will have to be given as well to the needs of the urban poor, whose access to health and other services is far worse than that of richer city dwellers and often not much better than rural conditions. Unmet need for family planning among the urban poor in Asia and sub-Saharan Africa, for example, is nearly as great as for rural populations (in South-east Asia it is greater). The urban poor are similarly disadvantaged with regard to skilled birth attendance and knowledge about avoiding HIV/AIDS.

Policy Developments since the ICPD

Nearly two thirds of developing countries responding to the 2003 UNFPA global survey (97 of 151) reported having taken some action on internal migration, compared to 41 per cent of respondents in 1994; 52 per cent have adopted plans on migration; 51 per cent have plans to influence spatial distribution of the population (including resettlement schemes, plans to redistribute population by creating new economic growth centres, and decentralization of social and economic planning and political decision-making); 16 per cent provide services to internally displaced persons; and 10 per cent have special institutions on migration. More-urbanized countries, and those with faster urban growth, were no more likely than other countries to have adopted multiple measures to address internal migration.

Three fourths of all governments, and nearly 80 per cent in developing countries, reported they were dissatisfied with the spatial distribution of their populations. The global survey indicates that a majority of developing countries have formulated policies on migration or allocated development investments with the aim of influencing population distribution. But the degree of attention given these issues varies widely.[6] Further development of policies addressing rural-urban movements and the conditions of life within cities will require more detailed data and research.

International Migration

According to the United Nations Population Division,[7] in 2000 there were 175 million international migrants in the world—1 in every 35 persons—up from 79 million in 1960.[8] Nearly 50 per cent were women, and 10.4 million were refugees. Between 1990 and 2000, two thirds of the growth in migrants took place in North America. Before 1980, the less-developed regions had a higher share of international migrants, but by 2000, three fifths were found in the more-developed regions.

Today, in an increasingly globalized economy, migration often provides employment opportunities, giving rise to an unprecedented flow of migrants, including increasing numbers of female migrants.[9] At the same time, there are growing numbers of refugees and people internally displaced by natural disasters, armed conflict, social unrest, or economic and political crises.

International migratory movements have big economic, sociocultural and demographic impacts on sending, transit and receiving areas.[10] Transit and receiving areas have had difficulties managing migration flows and integrating migrants into society. Sending areas have lost skilled labour and families have been divided, with women often becoming household heads after the departure of their husbands.

The migration of younger workers has left behind those too old for physical work in agriculture.

Heightened concerns about terrorism have prompted many countries to enhance security at their borders, leading to increased illegal immigration, particularly through smuggling and trafficking. Migratory movements have contributed to the spread of HIV/AIDS and other diseases.

The economic effects of migration run in both directions. Remittances from migrants flow from more- to less-developed countries. The World Bank reports that in 2002, total workers' remittances to developing countries amounted to $88 billion ($30 billion more than official development assistance), and that remittances flowing through official channels more than doubled between 1988 and 1999.[11]

The ICPD called on countries to: address the root causes of migration, especially those related to poverty, for instance, by promoting sustainable development to ensure a better economic balance between developed and developing countries, and defusing international and internal conflicts; encourage more cooperation and dialogue between countries of origin and countries of destination, to maximize the benefits of migration; and facilitate the reintegration of returning migrants.[12]

Recommendations included using short-term migration to improve the skills of nationals of countries of origin, collecting data on flows and numbers of international migrants and on factors causing migration, and strengthening international protection of and assistance to refugees and displaced persons.[13]

Echoing the ICPD and other international agreements, the Millennium Summit in 2000 agreed that countries should respect and protect the human rights of migrants, migrant workers and their families. The 2003 Final Report of the Commission on Human Security[14] stated, "The movements of people across borders reinforce the interdependence of countries and communities and enhance diversity".

Nevertheless, international migration remains a sensitive subject, and countries have not been able to agree to convene a United Nations conference to provide guidance to countries in addressing the issue, as some have proposed.[15]

Policy Response

The 2003 UNFPA global survey found that 73 per cent of developing countries responding (110 out of 151) reported having taken some action to deal with international migration, compared to 18 per cent in a similar inquiry in 1994. Nearly half of the countries had adopted programmes or strategies on international migrants or refugees; 37 per cent had enacted legislation on international migrants and migrant workers; 33 per cent had adopted a migration policy; 11 per cent had undertaken efforts to enforce international conventions on refugees, asylum-seekers and illegal migrants; and 10

per cent had passed laws on the trafficking of humans, especially women and children.

A growing number of countries have established coordination mechanisms of various types—across government agencies, between governments, and among governments, NGOs and international donors.

Policies in some African countries, like Ghana and the United Republic of Tanzania, emphasized settling refugees. In Latin American and Caribbean countries, the focus was on providing incentives for returning nationals, while the emphasis in Eastern Europe, the Arab States and Central Asian Republics was on protecting labour markets and combating drug trafficking.

To better integrate immigrants into their society, a few countries have adopted measures promoting equal opportunity in access to jobs, housing, health and education. Some developed countries have changed their family reunification policies in the past decade.

Since July 2002, for instance, Denmark no longer offers a statutory right to reunification with a spouse, and in most cases does not grant reunification if one of the spouses is younger than 24. New Zealand recognizes a wider range of family structures than it used to, but has also strengthened the legal responsibility of sponsors for the family members they bring into the country. Canada's policy, on the other hand, has

---

8 MAPPING PEOPLE AND THEIR NEEDS

Information systems developed in the past decade provide precise geographic information about people responding to demographic surveys. This will allow more detailed examination of the distribution of wealth, opportunities and challenges within countries, and better understanding of the push and pull factors driving population relocation and the impact of policies addressing it.

For example, recent research combines census and survey data to map the distribution of populations. The results suggest the land area covered by urban centres and their peri-urban settlements—and their impact on the environment—may be much

greater than earlier estimates based on administrative boundaries. The new methods also allow measurement of populations along coasts and in other ecologically sensitive regions.

This mapping has been used to study infant mortality in 10 West African countries. The results reaffirmed long-established findings (e.g., higher male mortality, urban advantage over rural areas, and the protective effects of mothers' education and improved sanitation), but also provided unexpected new insights into the high concentration of the poorest performing areas, which could lead to better targeting of programmes.

Significant investments in capacity building and technology transfer will be required to take full advantage of new data

collection technologies and analysis methodologies. More-detailed analyses of urban social networks and the characteristics of neighbourhoods also need to be incorporated into developing country research and programming.

This could facilitate more local development decision-making, and lead to better policies addressing the variety of settlement patterns with the aim of reducing poverty and improving the quality of life. For instance, mixed income communities may offer avenues for quicker advancement of the poor. Geographic targeting and use of local associations may help realize the ICPD vision of social participation.

become less restrictive, now including dependent children under 22 instead of 19.

A number of developed countries have introduced initiatives encouraging the immigration of skilled workers in response to labour shortages. Some have adopted policies aimed at attracting and retaining highly skilled students from developing countries.

To counter the growing trafficking of human beings, countries have tightened border controls and made asylum policies more restrictive; in some cases this has inadvertently made trafficking more profitable. In response, some countries have introduced severe penalties for human trafficking.

Although most receiving countries recognize the positive contributions of migration to the economic, social and cultural development of both migrant-receiving and migrant-sending countries, the growing levels of illegal immigration and the continuing flows of refugees and asylum-seekers remain major concerns.

Many countries favour more international cooperation to manage migration more effectively. Since 1994, eight regional and subregional consultation processes have been set up, covering nearly every country. The International Migration Policy Programme begun in 1998 has organized 15 regional meetings to promote cooperation and capacity building. And a Global Commission on International Migration was launched in December 2003; it is scheduled to issue recommendations to the UN Secretary-General in mid-2005.

5 Gender Equality and Women's Empowerment

The 1990s was an outstanding decade for bringing issues of reproductive health and rights, violence against women, and male responsibility for gender power relations to the centre of global and national debates on human rights and human development. The UN conferences of the 1990s, particularly the World Conference on Human Rights (Vienna, 1993), the ICPD (Cairo, 1994) and the Fourth World Conference on Women (Beijing, 1995), were central to a major paradigm shift in population policy.

In Cairo, the world's governments reached a consensus that affirmed their commitment to promote and protect the full enjoyment of human rights by all women throughout their life cycle. They also agreed to take action to accord more power to women and to equalize their relationships with men, in laws, economic systems and within the household.

The ICPD Programme of Action included, for the first time in a major international population policy document, a full and detailed chapter (Chapter IV) on women's empowerment and gender equality. In part, it stated that: ". . . improving the status of women also enhances their decision-making capacity at all levels in all spheres of life, especially in the area of sexuality and reproduction".

Gender equality and women's empowerment were at the heart of the ICPD vision. The Programme of Action's sexual and reproductive health and reproductive rights goals are strongly linked to, and mutually reinforcing of, its goals for women's empowerment and gender equality. The ICPD made a major new commitment in its objective ". . . to promote gender equality in all spheres of life, including family and community life, and to encourage and enable men to take responsibility for their sexual and reproductive behaviour and their social and family roles" (para. 4.25).

The ICPD also called on countries to "take full measures" to eliminate exploitation, abuse, harassment and violence against women, adolescents and children (para. 4.9). And it called for men to take

ACTIONS TO EMPOWER WOMEN

Countries should act to empower women and should take steps to eliminate inequalities between men and women . . . by: . . . eliminating all practices that discriminate against women; assisting women to establish and realize their rights, including those that relate to reproductive and sexual health; . . . eliminating violence against women; . . . eliminating discriminatory practices by employers against women such as those based on proof of contraceptive use or pregnancy status; . . . [and] making it possible through laws, regulations and other appropriate measures, for women to combine the roles of child-bearing, breastfeeding, and child rearing with participation in the workforce.

—from the ICPD Programme of Action, para. 4.4.

shared responsibility for parenting, valuing children of both sexes equally, educating them and preventing violence against them. It also urged actions to ensure that men actively participate with women in responsible behaviour in sexual and reproductive matters (para. 4.27).

In various countries, the paradigm shift of the ICPD also helped catalyse important changes in the approach of the UN system at the country level. For instance, in India, the ICPD approach has strong synergy with the UN Development Assistance Framework (UNDAF), which prioritizes gender equality and decentralization as crosscutting themes for all UN system assistance to India. The framework's main objectives in promoting gender equality are to enhance women's decision-making capability, to promote equal opportunity and to support policy change.

In India, the collaborative actions identified to promote gender equality are:

- Development of a gender policy analysis framework;

- Support for a comprehensive gender-disaggregated database;

- Support to promote gender equality;

- Assistance in developing gender-sensitive state plans;

- Promotion of inter-agency action research on gender.

In 2003, the Office of India's Registrar-General and Census Commissioner, the Ministry of Health and Family Welfare, and UNFPA drew attention to the problem of sex-selective abortion and female infanticide, and the resulting decline in the number of girls relative to boys, publishing a booklet entitled, *Missing: Mapping the Adverse Child Sex Ratio in India.*

In its 1999 review of ICPD implementation, the UN General Assembly called for redoubled action to redress gender inequalities, including the elimination of harmful practices, attitudes and discrimination against women and girls. Zero tolerance among the public was urged for son preference, unequal care for or valuing of girl children, and all forms of violence

directed against women—including female genital cutting, rape, incest, sexual violence and trafficking. Governments were encouraged to adopt legal changes, as well as encourage changes in the social, cultural and economic spheres.[1]

Achievements

Significant progress has been made in implementing these goals in the ten years since ICPD, but this progress has been uneven and still faces many challenges. UNFPA's 2003 global survey of national progress presents a mixed picture.[2] A number of countries have introduced laws and policies, but less has been done to translate these into programmes, implementation and monitoring.

Nonetheless, important steps have been taken.

For instance, in Mexico, the Women's Health Programme under the Secretary of Health has been training health sector employees to promote gender equity in their specific areas. Indonesia is implementing the President's Instructions on Gender Mainstreaming in National Development through regional and provincial development management teams that include government staff, local NGOs and researchers. Iran has established special centres for women police to provide services to women victims of violence, and prevention and counselling services including telephone hotlines.

In India, despite the continuing gender disparity in education, gender gaps in literacy appear to be diminishing in some of the states that traditionally have had the most serious problems, according to the 2001 Population Census. Innovative attempts are being made, as in the state of Haryana, to increase girls' school attendance by providing escorts to reduce families' concern about threats to their security. In Mexico, the National Population Council has initiated a major attempt to expand the scope of data collection on a broad range of issues related to sexual and reproductive behaviour.

NGOs, too, have undertaken a range of programmes to make real the ICPD's promise of gender equity and equality. In Calabar, Nigeria, for example, the Girls' Power Initiative mobilizes and empowers girls to take charge of their lives by opposing violence and demanding their rights. A corresponding pro-

TRAINING HEALTH WORKERS ON GENDER-BASED VIOLENCE

Gender-based violence is a worldwide problem that, studies show, may affect one out of three women. Abuses ranging from verbal abuse to rape to such traditional practices as female genital cutting are physically and psychologically damaging—and are human rights violations. Many victims are never seen by a medical professional to address their abuse, making assisting them a challenge.

As part of its work to counter gender-based violence, UNFPA has supported training of medical professionals, to make them more sensitive towards women who may have experienced violence and to meet their health needs. Pilot interventions have been tested in 10 countries—Cape Verde, Ecuador, Guatemala, Lebanon, Lithuania, Mozambique, Nepal, Romania, Russia and Sri Lanka.

Following consultations with health providers and clients, all women were screened for abuse in some pilot projects. Possible victims have been offered legal, medical and psychological support, and medical referrals when necessary. Some of the pilots have been undertaken with local authorities and hospitals, others work with NGO networks. Attention has been paid to involving communities, and to creating support networks for gender-based violence victims that include both police and healthcare providers, along with counselling services.

UNFPA has also held workshops for health providers on recognizing the effects of gender-based violence on women's health, and on how to detect and prevent abuse and assist victims. These have stressed the need for confidentiality and monitoring.

An evaluation found the pilots successful and worthy of continued support. Recommendations include a call for governments to recognize gender-based violence as a public health concern.

Based on this experience UNFPA has produced a manual, *A Practical Approach to Gender-based Violence*, which has been translated into seven languages.

---

gramme for boys trains them to become more gender-sensitive and supportive young men.

ADDRESSING INEQUALITY IN HEALTH PROGRAMMES. Many programmes to reduce unintended pregnancy work in settings where women have little autonomy and tend not to be assertive in their relationships with husbands or health care providers.

Interventions such as the Better Life Options programme for young women in India,[3] the Programme for Adolescent Mothers in Jamaica[4] and the Training of Trainers in Health and Empowerment in Mexico[5] aim to strengthen women's practical skills in long-term thinking, problem-solving and decision-making, and to persuade them that they are capable of making important decisions about their lives and health.

Some successful programmes educate women about reproductive and human rights; others offer training in literacy, employment skills, legal rights, parenting, child health, and social mobilization.

Global Survey Results

The 2003 UNFPA global survey provided a useful framework for assessing what has been done in the area of gender equality and women's empowerment. It covered five sets of measures: (i) to protect girls' and women's rights and promote empowerment; (ii) to address gender-based violence especially against girls and women; (iii) to improve access to primary and secondary education and address gender disparities in education; (iv) to instil attitudes favouring gender equality and support for women's rights and empowerment in boys and men; and (v) to promote male responsibility for their own and their partners' reproductive health.

PROMOTING GENDER EQUALITY. While more than half of the 151 responding developing countries had adopted national legislation, ratified UN conventions and established national commissions for women, similar progress was not made in formulating policies and putting programmes into place. Only one third of the countries had taken such action. Even fewer (only 13 countries) had developed advocacy programmes for gender equality.

MEASURES TO EMPOWER WOMEN. About half the countries had developed plans and strategies for women and to provide them with economic opportunities, but only 28 countries had increased women's political participation and just 16 had programmes to sensitize government officials.

GENDER-BASED VIOLENCE. The survey found that 91 countries had laws in place to counter and punish gender-based violence, but only 21 actually enforced these laws. Only 34 had trained service providers or government officials about gender-based violence, and only 33 had set up monitoring mechanisms.

ACCESS TO EDUCATION. Only 42 countries were able to increase public spending on schools, and only 28 provided incentives for poor families to send children to school. In addition, only 13 countries had incorporated gender sensitization into curricula, and only 16 had increased the number of girls' secondary schools.

MALE ATTITUDES AND RESPONSIBILITY. Only 20 countries reported developing youth and adolescent reproductive health education plans and programmes. Less than half the reporting countries had in place programmes to educate men about their own and their partners' reproductive health.

Legal Progress

Over the past 10 years, many countries have adopted new laws or amended legislation to advance gender equality, seeking to eliminate all forms of discrimination based on sex as well as to prevent gender-based violence and increase penalties for those who inflict it.

Among the countries adopting legislation to outlaw discrimination based on sex are Malta, Mauritius (this legislation also ensures equal rights for women regardless of their pregnancy or marital status) and Mexico. Colombia and Slovenia enacted laws to promote equal opportunities for women with men, and a decree in Costa Rica calls for improvements in the living conditions and opportunities of poor women.

Djibouti passed legislation adopting a National Strategy for the Improvement of Women in Development and a National Action Plan, which states that all policies and laws will be evaluated based on their impacts on the integration of gender into development. The law also details activities the Government will take to promote reproductive health and equal education for women, and to improve women's participation in decision-making (in the public sphere and the family) and in economic development. The Republic of Korea passed a law establishing a Commission on Gender Equality to manage policies on gender.

A number of constitutions, newly drafted or amended, contain strong provisions on gender equality. For example, Bahrain's 2002 constitution, while noting the Shari'a is the principal source for laws, affirms the equality of women and men in politics and in the economic, social and cultural spheres. Cuba's 2002 constitution affirms that spouses are equal in rights and duties. Timor-Leste's post-independence constitution affirms equal rights for women and men in marriage and the family and within social, economic and political life.

Rwanda's 2003 constitution also guarantees equal rights of spouses in marriage and divorce, outlaws discrimination based on sex, and establishes a National Human Rights Commission and a National Council of Women. It also guarantees the right of women and men to vote and run for office, calls for equal pay for equal work and establishes the right to education. In 2002, Togo amended its constitution to ensure gender equality before the law and in labour relations.

Poland has established a Plenipotentiary for the Equal Status of Women, located in the Prime Minister's office, to analyse women's legal and social status and promote equity through laws and policies.[6]

In 2000, in Azerbaijan, a presidential decree instructed the Government to ensure women and men are represented equally in the state administration and have equal opportunities under ongoing reforms. Government institutions were also directed to appoint a gender focal point in each district office.[7]

GENDER-BASED VIOLENCE. Addressing domestic and sexual violence directed against women is another priority for many governments' legislative action.[8] In Bangladesh, new laws make violence against women a punishable offence, and codes of conduct address sexual harassment in the workplace. Belgium, Peru and Yugoslavia have amended laws to define sexual harassment and make it a crime for which victims can sue and seek restitution.

Belgium, the Dominican Republic, Portugal, Spain and Uruguay, among others, have passed laws increasing penalties for gender-based violence. In Brazil, a 2003 law established a national, cost-free telephone hotline operated by specially trained staff for women to report domestic abuse.

Human trafficking has also been the subject of legal changes. Many countries enacted laws to combat trafficking of women and girls and many ratified international treaties.[9] The Democratic Republic of the Congo outlawed trafficking in children in its 2002 labour code.

While most governments say they recognize the importance of promoting gender equity and women's empowerment, many find it difficult to work directly with women at the community level. Accordingly, in countries such as Jamaica, Malaysia and Mozambique, women's NGOs are implementing such programmes. NGOs are often more effective in working with victims of gender-based violence, since they are perceived as being more sympathetic and are more likely to be trusted.

NGOs are also training police officers, judges and others in how to deal with victims of gender-based violence when they seek help. In Ethiopia, for example, the Association of Women's Lawyers is working against domestic violence and sexual abuse. Ethiopia's National Council on Traditional Practices and other NGOs are actively working to eradicate harmful traditional practices like female genital cutting. In the Philippines, NGOs have established women's crisis centres for victims of domestic violence.

Jamaican NGOs including the Association of Women's Organizations in Jamaica, Fathers' Incorporated and the Bureau of Women's Affairs collaborated between 1999 and 2002 to increase public awareness of gender-based violence. They worked to increase media coverage of the issue, and to educate police, the judiciary, and health and legal professionals about the importance of a strong response to violence against women and of support systems for victims.

REPRODUCTIVE RIGHTS. During the past decade, NGOs in many countries have become increasingly involved in monitoring reproductive rights and using the reporting procedures for international human rights instruments that their governments have ratified. Many submit "shadow reports" to complement those submitted by the government and attend sessions of the relevant monitoring committee when the report of their country is being examined.

In some countries such as India, Indonesia, Malaysia and Nigeria, human rights commissions can play an important role in ensuring that reproductive rights are observed and can provide redress in cases of violations. Other countries have ombudsmen or other mechanisms that civil society groups can use.

MALE INVOLVEMENT. NGOs are recognized as often being more effective than government agencies in encouraging men to take responsibility for their sexual and reproductive behaviour and their social and family roles. In Cambodia, for example, Men against Violence against Women actively participated with women's and other NGOs in a 16-day campaign on gender-based violence.

In the Philippines, NGOs are actively promoting male support for women's empowerment and rights with respect to reproductive health. And the Women's Centre of the Jamaica Foundation counsels young male parents and trains male peer educators through its programme, Young Men at Risk.

The ICPD and the MDGs

Millennium Development Goal 3, "Promote gender equality and empower women", reflects the ICPD's objectives. Its main target for measuring progress, though, is rather limited: eliminating gender disparities in primary and secondary education by 2005, and at all education levels by 2015.

The UN Millennium Project Task Force 3 on Primary Education and Gender Equality has adopted a broader operational framework that assesses gender equality and women's empowerment along three dimensions: (1) *human capabilities* as measured through education, health and nutrition; (2) *access to resources and opportunities*, referring to economic assets and political participation; and (3) *security* in terms of vulnerability to violence.[10]

Drawing on extensive evidence from a wide range of countries, the task force has identified strategic priorities for national and international action; the synergy between these priorities and the ICPD approach is strong. Central to strengthening women's capabilities are: elimination of gender gaps in secondary education, and increasing adolescents' and women's access to sexual and reproductive health information and services.

In 2003, WHO's Department of Gender and Women's Health provided an analysis of how looking through a gender lens would strengthen the effectiveness of efforts to realize the health-related MDGs in addition to Goal 3.

WHO's recommendations include, among others: paying attention to the impacts of son preference on the nutritional intakes and health care access for girls and women; reducing the work burdens of girls and women; addressing the growing feminization of the HIV pandemic due to gender-biased traditional practices and inappropriate programme emphasis; and addressing the specific effects of gender in the incidence of malaria and tuberculosis. To reduce indoor air pollution that particularly affects women and girls, WHO is putting priority on provision of cleaner cooking and lighting fuels.

Priorities for improving economic and political opportunity are: investing in infrastructure in women-friendly ways so that women's work time and drudgery levels are reduced; reducing discrimination against women in employment and earnings; reducing gender inequalities in property and inheritance; and increasing women's share of seats in local and national government bodies. Key to improving women's security is to take action that will significantly reduce the incidence of violence against girls and women.

Challenges: Filling the Half-empty Glass

A recent analysis of national reports on progress toward meeting the MDGs found that "Despite the rights-based perspective reflected by most reports in discussions on Goal 3 [gender equality and empowerment of women], the approach to women in discussions under other goals continues to be instrumental rather than rights-based. Examples are the discussions on child mortality in several reports, where women's lack of knowledge of care and feeding practices is most commonly identified as a barrier to achieving the goal. Such a formulation ignores the gendered variables that mediate child survival, while accepting without comment the invisibility of fathers in parenting and care."[11]

Even 10 years after Cairo, the report found that "women are still being seen in terms of their vulnerabilities" and cast most often within their traditional roles of mothers or as victims—not as actors in the development process.

Other impediments to progress include the continued lack of good quality data disaggregated by sex, the paucity of financial and technical resources for women's programmes at both international and national levels, and confusion about the relative merits of gender mainstreaming versus women-focused programmes.

DATA. Without sex-disaggregated data, it becomes impossible to put benchmarks on or monitor policy or programme effectiveness. In most countries, serious gaps still exist in available data on women's economic activity and decision-making ability and on the differential impacts of anti-poverty or other programmes. Data that allow cross-country comparisons are even more scarce. Efforts currently under way to fill these gaps need to be prioritized and strengthened.[12]

LIMITED RESOURCES. A second ongoing problem is the paucity of resources. Many commitments made by governments and agencies cannot be met because of the lack of funds. National programmes promoting women's advancement are particularly susceptible to arbitrary budget cuts in times of fiscal stringency.

For the rights-based approach to population issues to be translated into effective laws, policies and programmes, it is important that the most important existing human rights instrument, the Convention on the Elimination of All Forms of Discrimination Against Women (CEDAW), be used effectively.

The UN Millennium Project's Task Force 3 on Primary Education and Gender Equality has recommended using CEDAW to monitor progress and strengthen accountability on gender equality and women's empowerment. A 2000 study concluded that effective utilization of CEDAW depends on widespread knowledge of CEDAW; dialogue among government representatives, CEDAW committee members, and NGOs; use of sex-disaggregated indicators for tracking policies, laws, and budgets; and government recognition of how to link policies to CEDAW.

To date, CEDAW has been insufficiently used to track the implementation of the ICPD Programme of Action or to develop the mechanisms for such implementation.

A particularly unfortunate trend is the tendency to cut funding to women-focused programmes or agencies based on the argument that gender is now being mainstreamed throughout the institution.

MAINSTREAMING EFFORTS. Gender mainstreaming became the approach of choice in the 1990s in response to the recognition that women-focused programmes or agencies are easy to isolate or marginalize. However, gender mainstreaming is a difficult process and one in which good practices are still evolving.

The UNFPA global survey found considerable shortcomings in understanding of what a gender equity approach means and how to operationalize it within programmes and policies as called for by the ICPD. Mainstreaming efforts, undertaken without a women's agency to back it up, can be unfocused and even easier to ignore than women-focused initiatives. What is required is a combination of mainstreaming efforts (with a clear operations research approach to determine what works and what doesn't and why within key institutions[13]) and a well-funded and resourced women's machinery (including ministries for women's affairs or gender equality and focal points for gender issues within ministries, commissions and departments) that has the technical capacity and political placement to spearhead policies and programmes.

POLITICAL WILL. Behind these factors lie the questions of political clout and commitment. In situations where a vocal national women's movement is able to advocate for needed policies, programmes and resources, forward movement can be quick. In other situations, NGOs or other civil society organizations are running interesting programmes for gender equality, but most are not scaled up into government-supported programmes. However, even in situations where the women's movement is not strong, political leadership can play an important role in advocating for gender equality and women's empowerment at the policy level.

6 Reproductive Health and Family Planning

Gaps in reproductive and sexual health care account for nearly one fifth of the worldwide burden of illness and premature death, and one third of the illness and death among women of reproductive age.[1]

At the Cairo conference, the international community embraced a new, broad concept of reproductive health and rights, including family planning and sexual health. It called for integrating family planning and maternal and child health care within a wider set of services including the control of HIV and sexually transmitted infections (STIs). Expanding access to services and meeting clients' expressed needs and wishes were seen as essential to reducing unintended pregnancies, improving maternal health and curbing the HIV/AIDS pandemic.

The ICPD set the goal of ensuring universal voluntary access to a full range of reproductive health care information and services by 2015. Delegates also agreed that sexual and reproductive health is a human right, part of the general right to health.

There has been important progress since 1994 towards the ICPD goal of universal access to reproductive health services. The 2003 UNFPA global survey found that most countries have established or broadened reproductive health policies and programmes. Many have significantly changed the ways maternal and child health services and family planning are provided,[2] reorienting services to improve their quality and better meet clients' needs and desires. Rising contraceptive use among couples indicates greater

REPRODUCTIVE HEALTH—A HOLISTIC CONCEPT

Reproductive health is a state of complete physical, mental and social well-being, and not merely the absence of disease or infirmity, in all matters relating to the reproductive system and its functions and processes. Reproductive health therefore implies that people are able to have a satisfying and safe sex life and that they have the capability to reproduce and the freedom to decide if, when and how often to do so. Implicit in this last condition are the right of men and women to be informed and to have access to safe, effective, affordable and acceptable methods of family planning of their choice, as well as other methods of their choice for regulation of fertility which are not against the law, and the right of access to appropriate health care services that will enable women to go safely through pregnancy and childbirth and provide couples with the best chance of having a healthy infant.

—ICPD Programme of Action, para. 7.2

access to family planning.[3] Greater attention has been given to reproductive rights in laws and policies.

Since 1994, governments, international agencies and NGOs have focused on increasing the demand for contraception, not just expanding supply. Raising demand means better meeting clients' needs and desires, and offering a broad mix of family planning methods and higher quality care, with well-supplied facilities and well-trained, informative and courteous personnel.

But millions of people—particularly the poor—still lack access to quality services, including modern family planning methods, emergency obstetric services, and prevention and management of STIs. Demand is growing faster than resources to meet it. And in many countries, a variety of factors have constrained progress in integrating existing services and realizing the synergies envisioned by the ICPD.

GLOBAL SURVEY RESULTS. Country responses to the 2003 UNFPA global survey indicate progress in expanding access to reproductive health and in integrating family planning with other services. But they also reported considerable challenges. In the decade since the ICPD, countries have embraced the idea and the practice of reproductive health. They have broadened programmes to reach more people in need of services, and integrated family planning with pre- and post-natal care, delivery services, STI and HIV/AIDS prevention, cervical cancer screening, and referral for treatment where appropriate.

An increasing emphasis has also been placed on improving access for underserved groups, including the very poor and people living in rural areas, and on ensuring that the poor have a stronger voice in policy-making, so that information and services are adapted for their special circumstances. Countries are also integrating reproductive health, including family planning and sexual health, with primary health-care services in the context of health-sector reform and far-reaching changes in organizational, administrative and financing arrangements.

LEGAL DEVELOPMENTS. Since the ICPD, many countries have adopted laws to expand access for all women and men to reproductive health care and to ensure that

pregnant women and adolescents are not barred from or discriminated against in school or employment. The UNFPA global survey found that 46 countries had enacted new laws and legislation since 1994.

For example, in Papua New Guinea the requirement for a "husband's consent" for contraceptive use has been removed and adolescents over age 16 can access reproductive health services without parental consent. The Lao People's Democratic Republic has adopted a national birth-spacing policy, reversing a pro-natalist policy adopted in the 1990s. Under the

---

12 WHO ADOPTS GLOBAL STRATEGY ON REPRODUCTIVE HEALTH TO ACHIEVE MDGs

In May 2004, the 57th World Health Assembly adopted the World Health Organization's first strategy on reproductive health. The aim is to accelerate progress towards meeting the Millennium Development Goals and the reproductive health goals of the ICPD and its five-year follow-up.

The strategy identifies five priority aspects of reproductive and sexual health:

- Improving antenatal, delivery, postpartum and newborn care;

- Providing high-quality services for family planning, including infertility services;

- Eliminating unsafe abortion;

- Combating STIs, including HIV, reproductive tract infections, cervical cancer and other gynaecological morbidities;

- Promoting sexual health.

The Assembly recognized the ICPD Programme of Action, and urged countries to:

- Adopt and implement the new strategy as part of national efforts to achieve the MDGs;

- Make reproductive and sexual health an integral part of planning and budgeting;

- Strengthen health systems' capacities to provide universal access to reproductive and sexual health care, particularly maternal and neonatal health, with the participation of communities and NGOs;

- Ensure that implementation benefits the poor and other marginalized groups including adolescents and men;

- Include all aspects of reproductive and sexual health in national monitoring and reporting on progress towards the MDGs.

new policy contraception is provided for free and without coercion.

Belize's National Health Policy outlines reproductive rights, including voluntary counselling and testing for HIV infection; ensures tax exemption for NGOs in providing health services; and sets protocols for family planning services.

Twenty countries reported in the global survey on communications/advocacy campaigns that promote reproductive rights. Eighteen highlighted institutional changes. For example, in Micronesia, pregnant students are allowed to continue at school and adolescents are allowed access to contraception without written parental consent. Ten countries adopted family laws and legislation to make men more responsible for reproductive health.

Family Planning and Sexual Health

The ICPD called on all countries to take steps to meet the family planning needs of their populations and to provide, by 2015, universal access to a full range of safe and reliable family planning methods. The aims were to help couples and individuals meet their reproductive goals; to prevent unwanted pregnancies and reduce high-risk pregnancies; to make quality family planning services affordable, acceptable and accessible to all; to improve the quality of family planning information, counselling and services; and to increase men's participation in the practice of family planning.

BENEFITS OF CHOICE. Family planning enables individuals and couples to determine the number and spacing of their children—a recognized basic human right. Practical benefits are gained at many levels:

- To individuals, improved maternal and infant health; expanded opportunities for women's education, employment and social participation; reduced exposure to health risks; and reduced recourse to abortion;

- To families, reduced competition and dilution of resources; reductions in household poverty; and more possibility for shared decision-making;

- To the society, accelerated demographic transition; and the opportunity to use the "demographic

---

| 13 | **COSTS AND BENEFITS OF CONTRACEPTIVE USE** |

At a cost of about $7.1 billion a year, modern contraceptive use currently prevents annually:

- 187 million unintended pregnancies;
- 60 million unplanned births;
- 105 million induced abortions;
- 2.7 million infant deaths;
- 215,000 pregnancy related-deaths (including 79,000 from unsafe abortions);
- 685,000 children losing their mothers due to pregnancy-related deaths.

There are some 201 million women with unmet need for effective contraception. Meeting their needs, for an estimated annual cost of $3.9 billion, would avert some 52 million pregnancies each year (half of which would be delayed to a later time, in accordance with stated desires).

Preventing or delaying these unintended pregnancies would also prevent:

- 23 million unplanned births (a 72 per cent reduction);
- 22 million induced abortions (a 64 per cent reduction);
- 1.4 million infant deaths;
- 142,000 pregnancy related-deaths (including 53,000 from unsafe abortions);
- 505,000 children losing their mothers due to pregnancy-related deaths.

---

bonus" (see Chapter 2) to speed economic development.

Contraceptive Access and Use

Since 1994, family planning use has increased globally from 55 per cent of married couples to around 61 per cent; it has grown by at least 1 percentage point per year in 68 per cent of countries with available data and by at least 2 points per year in 15 per cent of these countries. Use varies regionally, ranging from about 25 per cent in Africa to nearly 65 per cent in Asia (where high use in China raises the average), and 70 per cent in Latin America and the Caribbean and in the developed regions.

However, many countries, particularly the poorest, still have restricted contraceptive access and choice.

When China (with a large population and high prevalence) is left out of the calculations, only 46 per cent of married women in Asia are using contraception. In the least-developed countries, the average is much lower.

Government support for methods of contraception—through government-run facilities, such as hospitals, clinics, health posts and health centres, and through government fieldworkers—has increased steadily since the 1970s. By 2001, the governments of 92 per cent of all countries supported family planning programmes.

Contraceptive prevalence has increased in each of 20 countries with two surveys since the mid-1990s, from a starting average of 28 per cent to 35 per cent. The proportion of desires being met by modern methods increased in 19 of these countries,[4] where there was an average 94 per cent increase among the poorest fifth of the population. In ten countries, the annual increase in met need for the poorest fifth was higher than the national average.

Unmet Need

Unmet need refers to women and couples who do not want another birth within the next two years, or ever, but are not using a method of contraception.[5] Unmet need results from growing demand, service delivery constraints, lack of support from communities and spouses, misinformation, financial costs and transportation restrictions.

The ICPD gave priority to reducing unmet need as a guiding principle in ensuring births by voluntary and informed choice.[6] The United Nations' 1999 review of implementation progress adopted benchmark indicators: reduction of unmet need to half of 1990 levels by 2005 and satisfaction of all stated fertility desires by 2015.[7]

The ability of people to implement their family size and spacing desires is a matter of great personal and demographic importance. In developing countries, total fertility regularly exceeds what people report as wanted fertility; on average the difference is around 0.8 children.[8] As family size desires shrink, unmet need tends to grow until service capacity catches up with the demand for fewer births and longer birth intervals. After that, further gains in service accessibility successively reduce unmet need.[9]

SUBSTANTIAL GAPS—AND GROWING. Despite the increase in contraceptive prevalence, some 137 million women still have an unmet need for contraception[10] and another 64 million are using traditional family planning methods that are less reliable than modern methods.[11] Overall, 29 per cent of women in developing countries have an unmet need for modern contraception. The highest proportion, several times the level of current use, is in sub-Saharan Africa where 46 per cent of women at risk of unintended pregnancy are using no method.[12]

Barriers to contraceptive use include:

- Lack of accessible services, and shortages of equipment, commodities and personnel;

- Lack of method choices appropriate to the situation of the woman and her family;[13]

- Lack of knowledge about the safety, effectiveness and availability of choices;

- Poor client-provider interaction;[14]

- Lack of community or spousal support;

- Misinformation and rumours;

- Side-effects for some, and insufficient follow-up to promote method switching or ensure proper use and dosage;

- Financial constraints.

Young people are particularly affected by a lack of temporary methods, inadequate confidentiality and privacy, and providers' lack of sensitivity to their perspectives.[15]

Despite decades of work to reduce these constraints, many problems persist. The decline of health systems in many countries has reduced access to services and the quality of personnel. Reduced donor support and inadequate national investments have hurt programmes. User fees, meant to promote sustainability and lower public sector costs, have increased inequities. Low salaries and poor working conditions lead employees to leave public health services for the private sector, further restricting access to the poor while driving the near-poor into poverty.

In 1999, at least 300 million married women lacked access to pills, IUDs or condoms.[16] Including voluntary sterilization, nearly 400 million lacked a full range of contraceptive choices. Since then, national reports show progressive improvements in the availability of multiple contraceptive methods.

Figure 3 shows the relationship between contraceptive use (including traditional methods) and unmet need, based on survey data from the past decade. The expected pattern is clearly evident: unmet need is highest in countries where prevalence is between 25 and 40 per cent.

Increases in modern method use generally reduce unmet need. At low levels of modern method use, unmet need varies greatly, depending on the level of unsatisfied demand and the degree to which people turn to available traditional methods before modern contraceptive services are available.

POVERTY AND DESIRED FAMILY SIZE. Poorer people tend to want more children than richer people.[17] Parents may see advantages in having more hands for subsistence farming, for example, or hope to ensure that they will have surviving children in their old age. Constrained opportunities also make the poor less likely to find social and economic incentives to invest more in fewer children rather than shallowly in many.[18]

But even in poor families, stated family size desires have been declining for decades, despite the persistence of attitudes and traditions favouring larger families. This is reflected in high levels of unmet need among the poor. There is much less difference between poorer and richer people with regard to wanting to postpone or avoid another birth than in contraceptive use.

The proportion of demand met by modern family planning rises steadily as modern use increases. In all cases, richer population groups are better able to satisfy their reproductive desires with modern contraception. Where prevalence is low, nearly a third of the couples in the richest income group who wish to delay or avoid a birth use modern contraception. Among the poor, the proportion satisfying their desires[19] with modern methods does not reach this level until contraceptive acceptance is more widespread.

Differences between poor and rich populations' access to family planning are staggering. In countries in sub-Saharan Africa, for example, women in the richest fifth of the population are five times more likely to have access to and use contraception than women in the poorest fifth.[20]

Figure 3: Unmet need and total prevalence

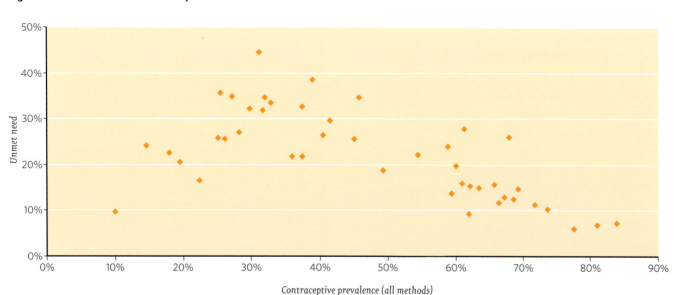

Source: Data provided by K. Johnson, ORC MACRO, International, from Demographic and Health Surveys.

Figure 4: Average total demand in wealth groups

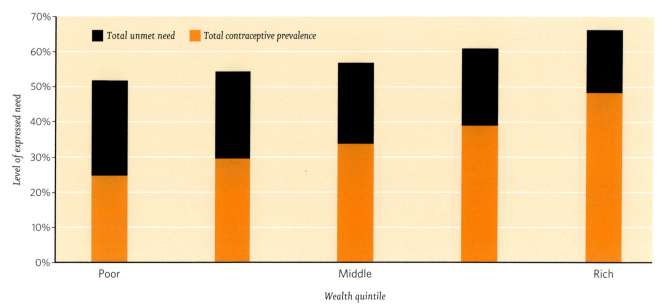

Legend: ■ Total unmet need ■ Total contraceptive prevalence

Y-axis: Level of expressed need
X-axis categories: Poor ... Middle ... Rich
Wealth quintile

Source: Data provided by K. Johnson, ORC MACRO, International, from Demographic and Health Surveys.

Choice of Methods

Modern methods today account for 90 per cent of contraceptive use worldwide. In particular, three female-oriented methods are most commonly used: female sterilization, intrauterine devices and pills. In one third of all countries, a single method, usually sterilization or the pill, accounts for at least half of all contraceptive use. Condom use has increased in the great majority of developing countries.

In the 2003 UNFPA global survey, 126 countries reported taking measures to expand contraceptive choice. Eighty-eight have taken multiple steps, including increasing the availability of emergency contraception (68 countries) and female condoms (65), improving logistics for ensuring contraceptive availability (43), and providing subsidized or free contraceptives or services (27).[21]

Emergency contraception refers to the prevention of pregnancy after unprotected sexual intercourse. Research over the past 30 years has shown that emergency contraceptive pills (special doses of ordinary contraceptive pills) are safe and effective when used within 72 hours. As stated by WHO, "Emergency contraceptive pills do not interrupt pregnancy and thus are no form of abortion."[22]

A growing number of countries have introduced emergency contraception since the ICPD; some have made it easier for women to access it, for example, by ending restrictions on over-the-counter sales. India,

---

14   UNFPA HELPS COUNTRIES EXPAND METHOD MIX

UNFPA supplies 40 per cent of the contraceptive commodities provided by the international donor community. It also is an important partner of governments and donors in responding to supply problems.

In 2003, 60 countries reported on service delivery points offering multiple methods of contraception, both for UNFPA programme areas and for the nation as a whole. In 24 countries, the proportion of facilities offering at least three methods was higher in UNFPA areas, and in 24 countries the service offerings were identical. In 11 other countries where the Fund focuses specifically on improving access for underserved locations (usually the poorest areas) or groups (especially adolescents), a smaller proportion of UNFPA programme sites offer at least three methods.

Twenty countries have reached and maintained universal access to at least three contraceptive methods at service delivery points in UNFPA programme areas. Reports from 26 countries show successive expansion of contraceptive choice. Access has declined in only five countries, in three of them after civil conflict.

Figure 5: Percentage of family planning demand satisfied in different wealth groups at three levels of modern contraceptive prevalence

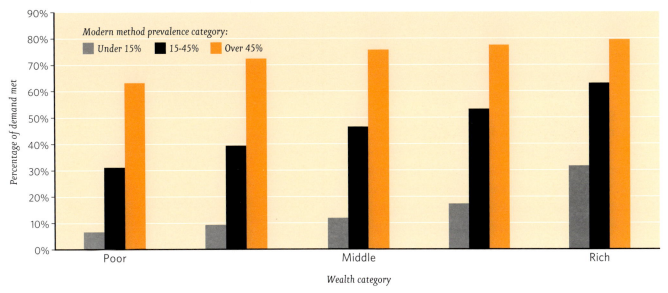

Source: Data provided by K. Johnson, ORC MACRO, International, from Demographic and Health Surveys.

Iran and Nepal provide it through the national family planning programme. In the Dominican Republic, emergency contraception can be obtained through private pharmacies, while in Malaysia and Pakistan, NGOs are supplying it.

Programmes addressing sexual violence often offer emergency contraceptive pills along with counselling to women who have been raped. In Chile, doctors and emergency rooms can distribute the pills to women who have been raped.

Sexually Transmitted Infections

Some 340 million new cases of sexually transmitted bacterial infections occur each year (including syphilis, gonorrhoea, chlamydia and trichomoniasis) in people between ages 15 and 49.[23] While most are treatable, many are undiagnosed and go untreated because of the lack of accessible services.[24] Untreated STIs are a leading cause of infertility, which affects 60 to 80 million couples worldwide.

More than three fourths of the 5 million new cases of HIV infection each year are sexually transmitted (see Chapter 8). An STI that goes untreated can increase the risk of HIV infection and transmission by up to 10 times.[25] Another viral STI, the human papillomarvirus, is linked to cervical cancer, which

kills 240,000 women each year. Condoms, both male and female, are the only contraceptive methods that provide significant protection against viral and bacterial STIs.[26]

The ICPD called for screening and treatment for STIs, along with information and counselling, to become "integral components of all reproductive and sexual health services".[27]

GENDER DIFFERENCES. Over the past decade, increasing attention has been given to the critical role of gender in diagnosing and treating STIs. Women are at greater risk of infection than men, and screening is more difficult: 70 per cent of women with STIs do not have symptoms (compared to 10 per cent of men).[28]

The management of STIs can have an important place in the provision of family planning. Where equipment is not available to test for STIs, health workers use a "syndromic approach" to diagnosis, based on risk factors and client symptoms. But this approach has limitations, and infections often go undiagnosed and untreated.[29]

Untreated STIs in pregnant women can facilitate the transmission of infection to the infant and raise the risk of a pre-term or low birth weight delivery or infant blindness.[30]

INTEGRATION. Providing STI screening, diagnosis and treatment as part of reproductive health care offers the opportunity to reach millions of women who seek such services, many of whom have no other contact with the health care system. Experience shows that integrating STI prevention, family planning, and counselling on sexuality and partner relationships can result in greater use of services.[31]

Integrated services, rather than stand-alone facilities or treatment by private doctors, allow savings in costs, staff, supplies and equipment, and are typically more convenient for clients.[32] But efforts to dismantle vertical STI programmes have met some resistance.

Indonesia began integrating STI services with other reproductive health services in 1995. To overcome the stigma associated with STI care or concerns about sex workers using health facilities, STI services were offered outside normal clinic operating hours and in separate examination rooms.[33]

In the 2003 UNFPA global survey, 43 countries reported taking measures since the ICPD to integrate information on STIs and/or HIV/AIDS prevention into primary health care.[34] Ecuador, Liberia, Mozambique and Zimbabwe, for example, now include STI services in primary health facilities.[35]

Quality of Care

The ICPD Programme of Action recognized that in addition to making reproductive health services universal, "family planning programmes must make significant efforts to improve quality of care" (para. 7.23). The aim should be to "ensure informed choices and make available a full range of safe and effective methods" (para. 7.12).

Since 1994, services in many countries have been reoriented to improve their quality and better meet clients' needs and wishes—through a wider choice of contraceptive methods, better follow-up and improved training of staff to provide information and counselling (with an emphasis on sensitivity, privacy, confidentiality and informed choice). Improving services for poor populations is another global priority.[36]

The publication in 1990 of a quality of care framework[37] established the components of good reproductive health care. Clients need a choice of contraceptive methods, accurate and complete infor-

---

15 | WHAT CLIENTS CONSIDER QUALITY CARE

Clients assess the quality of the services they receive. If given a choice, they will use facilities and providers that offer the best care as they perceive it. Studies around the world suggest that clients want:

- Respect, friendliness and courtesy;

- Confidentiality and privacy;

- Providers who understand each client's situation and needs;

- Complete and accurate information, including full disclosure about contraceptives' side-effects;

- Technical competence;

- Continuous access to supplies and services that are reliable, affordable and without barriers;

- Fairness. Information and services should be offered to everyone regardless of age, marital status, sex, sexual orientation, class or ethnicity;

- Results. Clients are frustrated when they are told to wait or come back.

---

mation, technically competent care, good interaction with providers, continuity of care, and a constellation of related services. Another framework detailed the support, tools and resources that providers need to offer quality care.[38]

Efforts to improve quality focus on improving the service environment to meet clients' needs by involving all levels of staff in identifying problems and suggesting solutions. After the ICPD, approaches that were already widely used in developed countries were translated for use in international family planning programmes.[39]

CONCRETE ACTIONS. The 2003 UNFPA global survey found that 143 countries had taken steps to improve access to quality reproductive health services, with 115 reporting multiple actions. These include increasing staff and training (77 countries); introducing quality standards (45), and improvements in management and logistics (36). In Bangladesh, the Democratic People's Republic of Korea and Mongolia, for example, protocols and quality control measures are now in place for a wide range of reproductive health services. Indonesia is updating existing protocols. Jamaica is establishing indicators for assessing and monitoring the quality of care.

Public sector, family planning association and women's health NGO programmes in Guatemala, India and Kenya all include providing quality care as part of their goals and objectives.[40]

IMPACTS OF QUALITY. Quality care can increase demand for services by helping clients select an appropriate contraceptive method and continue using family planning if they wish to limit or space their pregnancies. Women and men in communities with quality services are more likely to use family planning than those who are not, as a study in Peru showed.[41] In rural areas of the United Republic of Tanzania, perceptions of a family planning facility's quality of care have a significant impact on community members' contraceptive use.[42]

Being able to choose a contraceptive method matters to clients. In Indonesia, 91 per cent of women who were given the method they wanted continued to use it after one year, compared to 38 per cent of those who had not been given their method of choice.[43] In Gambia and the Niger, new users who received good counselling on side-effects were one third to half as likely to discontinue contraceptive use after eight months as those who perceived such counselling to be inadequate.[44]

EMPOWERING THE POOR. Better treatment particularly makes a difference to poor women. In a recent Bangladesh study, women who felt they had received good care[45] from fieldworkers were 60 per cent more likely to adopt contraception and 34 per cent more likely to continue its use than those who perceived they had received poor care.[46] While service quality affected contraceptive adoption for all women, it was far more important as a determinant for continued use among poor and uneducated women.

TRAINING EFFORTS. Interaction between clients and providers is critical to good care. Providers need to explore clients' thinking about health decisions, address their concerns about side-effects and encourage them to play an active role in consultations. Providers' knowledge and interpersonal skills can be improved by defining clear expectations for interaction with clients, giving feedback on their performance and making training more effective. It is also important to provide adequate compensation, space, supplies and time; and to match workers with jobs for which they have the skills.[47]

Countries as diverse as Senegal, Turkey and the United Republic of Tanzania have, since the ICPD, undertaken system-wide reforms to provide quality care to clients. They have strengthened training, expanded educational activities, upgraded infrastructure and equipment, updated policies and procedural guidelines, and strengthened management systems.[48]

Many other countries have strengthened staff training and supervision and improved method availability and choice.[49]

Many countries have worked to upgrade their reproductive health facilities. Measures taken include: certification or accreditation of facilities (Mozambique); strengthening infrastructure and ensuring that specialized follow-up care is available (Brazil); piloting mobile health units (Armenia and El Salvador); and providing free or low-cost services for slums and urban squatter settlements (148 countries).

16	**PROMOTING PARTICIPATION**

The ICPD stressed the importance of involving the beneficiaries of reproductive health programmes in planning, implementation and monitoring. The 2003 UNFPA global survey found that 124 countries reported having taken key measures in this area, with 48 reporting multiple measures. Some have conducted public hearings or consumer surveys and involved communities in developing programmes that reflect the needs and opinions of the population.

Kenya has included village chiefs and traditional healers as community resource persons. Malaysia has organized dialogues between service providers and clients. Brazil has set up national, regional and municipal health councils. Honduras has used questionnaires, focus groups and in-depth interviews to elicit feedback on all health systems. Latvia has created a "Patients Rights Bureau" that conducts surveys of patient satisfaction with health care.

A number of donors and international organizations have launched activities to promote civil society participation in meeting reproductive health needs. Thirty-four national family planning associations of the International Planned Parenthood Federation are undertaking a five-year initiative to identify and correct deficiencies in quality.

Stronger Voices for Reproductive Health

Since the ICPD, UNFPA has worked to improve the quality of care, increase access to services, ensure adequate supplies and equipment, and upgrade the technical, managerial and interpersonal skills of health staff by providing technical support, equipment and training.

More recently, attention has also focused on mobilizing communities to push for higher quality health services and more participation of women in their management. UNFPA is supporting "Stronger Voices for Reproductive Health", an innovative initiative that aims to empower users with knowledge about their reproductive health and rights and supporting community mechanisms so they will have a "stronger voice" to ensure steps are taken to improve reproductive health care. UNFPA partners with the International Labour Organization, UNICEF and WHO in this initiative.

COMMUNITIES OF EDUCATED CLIENTS. The premise is that better-informed users will have improved interactions with providers and will also be more likely to mobilize for change at the community level. Promoting the collective action of communities to demand quality care can also help ensure that the decentralization of health services under way in many countries does not result in a reduction of resources for reproductive health care.

The Stronger Voices project has built bridges between organizations that have previously not worked together—reproductive rights advocates, organized women's groups, health care providers and organizations focusing on community financing or health reform.

Stronger Voices started in India, Kyrgyzstan, Mauritania, Nepal, Peru and the United Republic of Tanzania. Activities included participatory approaches to increase women's access to reproductive health services, linking women's groups with providers to promote better care and mobilizing young people to work with providers on youth-friendly services.

In Kyrgyzstan, community groups raised funds and renovated field obstetric stations; community women are speaking out against the old tradition of "bride-napping" as a violation of reproductive rights. In Mauritania, two community-based micro health

insurance schemes were created to ensure poor women have access to maternity services. Project partners in Nepal developed a groundbreaking National Strategy for Quality of Care for Reproductive Health Services, which incorporates "demand" and reproductive rights as essential to delivering good quality services.

In the United Republic of Tanzania, project partners have established the country's first rights-based approach to quality of care, with an emphasis on local capacity building and community monitoring of services that dovetails with decentralization efforts.

PAYING FOR SERVICES. Many countries are assessing means of charging for services, given shortfalls in government and donor funding for reproductive health. Cost-recovery efforts have had mixed results. In Bangladesh, for example, the 1997 Health and Population Sector Strategy prompted NGOs to move away from home-based provision of family planning and to charge modest fees for services that were previously free. At the same time, the NGOs worked to upgrade their services.

An assessment of the changes found that clients appreciated the services provided and felt they were treated with kindness and respect in the NGO clinics. It also found, however, that the changes resulted in "a widespread feeling that poor people face discrimination in health facilities and quality services are out of their reach because they cannot pay for them".[50]

Securing the Supplies

The ability to formulate and provide quality reproductive health depends on having in place political support, funding, people, facilities and commodities.

While national programmes often have to cope with adversity—such as scarcities of personnel and facilities, a lack of running water, regular power outages and disruptions of transport—the absence of commodities means that investments and effort will be largely wasted.

"Reproductive health commodity security" involves assuring an adequate and secure supply of essential reproductive health supplies. These commodities must be repeatedly procured, delivered and distributed to where they are needed when they are needed. In many poor countries, reproductive health programmes will depend heavily upon donor assistance for commodities for a long time.

Over the past ten years, donor support for reproductive health supplies, including contraceptives for family planning and condoms for AIDS prevention, has declined, creating a growing gap between generally accepted estimates of need and what is being supplied.

In the early 1990s, just four international donors provided some 41 per cent of overall estimated requirements for contraceptives—pills, intra-uterine devices, injectable contraceptives and condoms. (Systems to accurately quantify the supply and demand for other reproductive health commodities are still under development.) The United States Agency for International Development (USAID), which had dominated public sector contraceptive supply since the 1960s, was the largest, accounting for almost three fourths of the $79 million in reported donor support for 1990.[51]

By 2000, the number of active donors had grown to 12 or more, but total donor support (adjusting for inflation) remained relatively flat during the decade. USAID's share fell to 30 per cent, while share provided by UNFPA grew to 40 per cent. These agencies and four others (Population Services International, the World Bank, the German Federal Ministry for Economic Development Cooperation and the United Kingdom's Department for International Development) accounted for 95 per cent of contraceptive commodities provided to developing countries.

In 2001, the Netherlands, the United Kingdom and Canada responded to serious supply shortages in some countries by contributing an additional $97 million to UNFPA for commodities and technical support to strengthen national capacity and improve access. The $224 million in total donor support that year represented an increase of almost 50 per cent over the previous year, but in 2002 (the latest year for which figures are available), the total dropped back to $198 million.

To meet the same 41 per cent share of contraceptive and condom supply needs that donors provided in 1990, their support would need to be around $450 million in 2004. Considerably more would be needed to meet all of the overall projected reproductive health commodity costs and to improve service delivery.

It is unlikely that developing country governments, NGOs and commercial sectors will be able to make up for the lack of growth in donor support for reproductive health commodities. As a result, we can expect commodity shortfalls and disruptions of reproductive health services with grave consequences for the health of women and children.[52]

On top of growing requirements for commodities support, developing countries need both technical support and funding to increase national health programmes' human, financial and technical capacities to collect, analyse, report and properly use data on reproductive health supply and demand; and to secure, store, and distribute the necessary supplies.

UNFPA'S ROLE AND PRIORITIES. UNFPA leads the global effort to ensure an adequate and steady flow of reproductive health supplies including contraceptives. The Fund is the largest international provider of such supplies, and the only provider for some 25 countries. In 2001 and 2002, it received supply requests from 94 countries totalling $300 million.

UNFPA also helps countries plan for their needs, undertakes advocacy to mobilize stable financing, works with governments and other partners to strengthen national capacity, coordinates partners' efforts, and collects data on donor efforts to facilitate cooperation and assure accountability.[53]

In 1999, in collaboration with NGO partners, UNFPA began work to develop a global strategy for securing reproductive health supplies.[54] Two important partnership mechanisms have been developed, the Supply Initiative (SI) and the Reproductive Health Supply Coalition.

17	**CONSEQUENCES OF THE FUNDING GAP**

Each $1 million shortfall in contraceptive commodity assistance will result in an estimated:

- 360,000 unintended pregnancies;
- 150,000 induced abortions;
- 800 maternal deaths;
- 11,000 infant deaths;
- 14,000 deaths of children under five.

With funding from the Bill & Melinda Gates Foundation and the Wallace Global Fund, the Supply Initiative has established a web-based information system to consolidate procurement data from UNFPA, USAID and the International Planned Parenthood Federation, and eventually from other donors. In the future, it will forecast each country's supply needs.

The Reproductive Health Supply Coalition, a diverse partnership, is exploring the possibility of forming a new mechanism to help mobilize resources and promote collaboration. But so far, donors have not shown enough interest to justify such a move.

To strengthen national capacity, UNFPA recently facilitated half a dozen regional workshops where participants—UNFPA field staff and government representatives—developed model plans for the management of reproductive health supplies.

OTHER INITIATIVES. The World Health Organization and UNFPA recently issued a joint draft discussion document titled, "Essential drugs and other commodities for reproductive health services". Intended in part to ensure a common understanding of the term "reproductive health commodities", the document draws upon the essential medicines concept introduced by WHO in 1977 and lists commodities needed at the primary health care level (for family planning, maternal and neonatal health, and prevention of reproductive tract infections and HIV) as well as products needed for maternal care at the first referral level. It recognizes four enabling factors needed to ensure sustainable access to these crucial items of care:

• Rational selection based on a national essential drugs list and evidence-based treatment guidelines;

• Affordable prices for governments, health care providers and consumers;

• Sustainable financing through equitable funding mechanisms such as government revenues or social health insurance;

• Reliable supply systems incorporating a mix of public and private supply services.[55]

In some developing countries, management information systems are providing reliable logistics data for forecasting, procuring and distributing supplies.

THE ROAD AHEAD. Between 2000 and 2015, contraceptive users in developing countries are expected to increase by 40 per cent as the number of reproductive age couples grows by 23 per cent and demand for family planning becomes more widespread.[56] UNFPA has projected contraceptive commodity requirements in 2015 at about $1.8 billion, of which $739 million could be expected to come from donors based on 1990 support levels. These figures include condoms for HIV/AIDS and STI prevention.

Achieving this level of needed support will require: strengthened political leadership in both donor and recipient countries; better advocacy to generate long-term financial support; cost-recovery mechanisms, where appropriate; more effective coordination among the main international partners; new mechanisms in developing countries for planning and monitoring supply use; more reliable, country-generated data; and better accountability on the part of all partners.

Men and Reproductive Health

Some of the most innovative work since the ICPD has aimed to involve men in protecting their partners' reproductive health as well as their own. Through their greater access to resources and power, men often determine the timing and conditions of sexual relations, family size and access to health care. Reproductive health programmes are increasingly being designed to counter the ways that gender inequality limits women's—and sometimes men's—access to health care.[57]

Prior to the ICPD, the population field tended to focus almost exclusively on the fertility behaviour of women, paying little attention to men's roles in its study of the macro dimensions and implications of population growth and fertility rates.[58] As a consequence, basic family planning programmes served women almost exclusively.

ATTITUDES TOWARD INFORMATION AND SERVICES. Research has long shown that men want to know more about reproductive health and want to support their partners more actively.

Men's desire to limit their family size often makes it possible for women who want to use contraception to do so. Research on male attitudes and practices, including a 17-country analysis of Demographic and Health Survey data collected on husbands during the 1990s,[59] suggests that views of men and women on contraception and family size are much closer than many in the field once believed.[60] Men generally want more and better information and access to services.[61] Those aged 15-24 want fewer children than men 25-34, who in turn want fewer than men in their 50s.[62]

When programmes exclude men, they undermine their own effectiveness. Men's reproductive health directly affects that of their partners, a reality that the AIDS pandemic has brought sharply into focus. Treating sexually transmitted infections in women makes little sense unless the partners who infected them are also treated and involved in prevention education.

INCLUDING MEN IN REPRODUCTIVE HEALTH. In virtually every country, and in thousands of governmental and NGO programmes, creative ways are being found to draw men into reproductive health programmes. Men have responded positively to these efforts.

All countries replying to the 2003 UNFPA global survey reported taking measures to promote male contraceptive methods. Education campaigns on men supporting women have been carried out in all of the Central Asian countries and in most countries in Asia and the Pacific, Africa and the Caribbean. More than half of the Caribbean countries reported developing plans to encourage more male involvement in reproductive health.

DIVERSE APPROACHES TO "MALE INVOLVEMENT". Programmes developed in recent years have taken different approaches to involving men in reproductive health. One approach focuses on men as obstacles to women's contraceptive use and as an untapped group of potential users themselves.[63]

A second group of programmes emphasize the need to provide men with sexual and reproductive health care, remedying their traditional exclusion from such services.[64] Programmes can improve men's access to sexual and reproductive health services by making existing services more receptive: welcoming men,

both as clients and as supportive partners or fathers, retraining staff, providing information and services for men, hiring and training male counsellors, and even altering clinic décor. In trying to make men welcome, programmes need to recruit and train male health workers, who can be important advocates and role models for healthful behaviours and supportive partnerships.

WORKING TO CHANGE GENDER NORMS. A third approach focuses on men as supportive partners of women and seeks opportunities to address the ways that social positions constrain the sexual and reproductive roles of women and men. Some programmes explicitly address inequitable gender norms that harm the health of both men and women. They work to educate men about the ways in which control over family resources, violence at home, or views of male or female sexuality, for example, can inhibit good reproductive health.

Programmes such as Mobilizing Young Men to Care in South Africa, the Better Life Options Programme for Boys in India, the Men Can Stop Rape's "Strength Campaign" in the United States, the Conscientizing Male Adolescents Programme in Nigeria, and *Cantera* in Central America seek to transform the values that underlie harmful behaviours.[65] They have shown that encouraging men to discuss their beliefs leads them to question harmful elements of traditional masculinity and that men welcome the opportunity to do things differently.

Some programmes promote communication and respect between men and women on reproductive health issues, and aim to build the negotiating skills of both sexes. Many efforts focus on young and unmarried men whose ideas about gender roles and sexuality are still evolving.[66] From Costa Rica to Kenya to the Philippines, programmes are working with young men to address their sexual health choices and to develop their skills. The Mathare Youth Sports Association in Kenya, for example, has established a peer education programme for HIV prevention.[67] The Brazilian NGO, ECOS, emphasizes the links between masculinity, fatherhood and health.[68]

Working with men in leadership positions who can influence other men and advocate for women's health is another important strategy.

7   Maternal Health

Obstetric complications are the leading cause of death for women of reproductive age in developing countries today, and constitute one of the world's most urgent and intractable health problems.[1] Reducing maternal death and illness is recognized as a moral and human rights imperative as well as a crucial international development priority, including by the ICPD Programme of Action and the Millennium Development Goals.

Tragically, despite progress in some countries, the global number of deaths per year—estimated at 529,000, or one every minute—has not changed significantly since the ICPD, according to recent estimates by WHO, UNICEF and UNFPA;[2] 99 per cent of these deaths occur in developing countries. Millions more women survive but suffer from illness and disability related to pregnancy and childbirth. Although data are hard to come by, the Safe Motherhood Initiative, a coalition of UN agencies and NGOs, estimates that 30 to 50 morbidities—temporary as well as chronic conditions—occur for each maternal death.[3]

New Approaches

Recognizing that most of these deaths and injuries could be prevented with wider access to skilled care before, during and after pregnancy, the ICPD called on countries to expand maternal health services in the context of primary health care and develop strategies to overcome the underlying causes of maternal death and illness.[4]

Over the past 10 years, global priorities for reducing maternal death and illness have undergone a paradigm shift. In the past, researchers and practitioners thought that high-risk pregnancies could be detected and treated and that antenatal care could prevent many maternal deaths. They also called for training of traditional birth attendants (TBAs) to reduce risks of death or illness during pregnancy.

However, these two interventions did not reduce maternal mortality.[5] Broad agreement now exists among health professionals and policy makers that most maternal deaths stem from problems that are hard to detect or screen for—any woman can experience complications during pregnancy, childbirth and the post-partum period—but are almost always treatable, provided quality emergency obstetric care is accessible.

Since the mid-1990s, governments, international agencies, including UNFPA, researchers and civil society have focused on the interventions judged to be the most effective: expanding women's access to skilled attendance at delivery; improving facilities for and women's access to emergency obstetric care to treat pregnancy complications; and ensuring that referral and transport systems are in place so women with complications can receive needed care quickly.

Also high on the list of global safe motherhood priorities are making sure women have access to family planning services to reduce unwanted pregnancies; and improving the overall quality and capacity of countries' health systems, especially at the district level; and strengthening human resources.

A FOCUS ON RIGHTS. An additional feature of the post-ICPD period is the recognition that maternal deaths and disability are violations of women's human rights, and are strongly tied to women's status in society and economic independence.[6] Various human rights conventions support the view that women have a right to health care that enhances the likelihood that they survive pregnancy and childbirth.[7] Rights-related issues like the role of gender

MATERNAL HEALTH SERVICES

[Maternal health] services, based on the concept of informed choice, should include education on safe motherhood, prenatal care that is focused and effective, maternal nutrition programmes, adequate delivery assistance that avoids excessive recourse to Caesarian sections and provides for obstetric emergencies; referral services for pregnancy, childbirth and abortion complications; post-natal care and family planning....

—from the ICPD Programme of Action, para. 8.22.

inequalities in maternal health and the impact of gender-based violence on pregnancies are receiving greater attention at all levels.[8]

Still, ten years after Cairo, women's needs often do not rank high on governments' or communities' lists of priorities. Women still lack full power to choose the obstetric care they want. Poverty, conflict and natural disasters worsen reproductive health and add new challenges to ensuring safe motherhood.[9]

The ICPD set a goal of reducing maternal mortality to one half of the 1990 levels by 2000 and a further one-half reduction by 2015. Countries were also urged to reduce the differences between developing and developed countries and within countries, and to reduce greatly the number of deaths and morbidity from unsafe abortion.

The 1999 review of ICPD implementation stressed the connection between high levels of maternal mortality and poverty, and called on states to "promote the reduction of maternal mortality and morbidity as a public health priority and reproductive rights concern" by ensuring that "women have ready access to essential obstetric care, well-equipped and adequately staffed maternal health-care services, skilled attendance at delivery, emergency obstetric care, effective referral and transport to higher levels of care when necessary".[10]

The Millennium Summit in 2000 also identified maternal health as an urgent priority in the fight against poverty. Millennium Development Goal 5 calls for a 75 per cent reduction by 2015 in the maternal mortality ratio (the number of maternal deaths for every 100,000 births) from 1990 levels.

Meeting these goals will be difficult. In the developing world as a whole, approximately 65 per cent of all pregnant women receive at least some care during pregnancy; 40 per cent of deliveries take place in health facilities; and skilled personnel assist slightly

Table 1: Maternal mortality estimates by region, 2000

Region	Maternal Mortality Ratio (Maternal Deaths per 100,000 Live Births)	Number of Maternal Deaths	Lifetime Risk of Maternal Death, 1 in:
WORLD TOTAL	400	529,000	74
DEVELOPED REGIONS	20	2,500	2,800
Europe	24	1,700	2,400
DEVELOPING REGIONS	440	527,000	61
Africa	830	251,000	20
Northern Africa	130	4,600	210
Sub-Saharan Africa	920	247,000	16
Asia	330	253,000	94
Eastern Asia	55	11,000	840
South-central Asia	520	207,000	46
South-eastern Asia	210	25,000	140
Western Asia	190	9,800	120
Latin America & the Caribbean	190	22,000	160
Oceania	240	530	83

Source: WHO, UNICEF, and UNFPA, 2003, *Maternal Mortality in 2000: Estimates Developed by WHO, UNICEF, and UNFPA.* Geneva: World Health Organization.

more than half of all deliveries. But just 35 per cent of deliveries in South Asia were attended by a skilled attendant in 2000; in sub-Saharan Africa it was 41 per cent (up from one third in 1985); in East Asia and in Latin America and the Caribbean, the proportion was 80 per cent.[11]

In many settings, available safe motherhood services cannot meet demand or are not accessible to women because of distance, cost or socio-economic factors. Pregnancy care may be consigned to a low place on household lists of priorities given its costs in time and money. Too many women are still seen as not worth the investment, with tragic consequences for them, their children, who are less likely to survive or thrive without a mother, and their communities and countries.

GLOBAL SURVEY RESULTS. In their responses to the 2003 UNFPA global survey, 144 countries reported having taken specific measures to reduce maternal deaths and injury; 113 reported multiple measures. The most common were training health care providers (76 countries); instituting plans, programmes or strategies (68), improving ante- and post-natal care (66), upgrading data collection and record keeping (45), and providing information or advocacy (40).

But only some countries have been successful in reducing maternal mortality (most are middle income; a few are poor). In China, Egypt, Honduras, Indonesia, Jamaica, Jordan, Mexico, Mongolia, Sri Lanka and Tunisia, deaths have been reduced significantly over the past decade. Common to all these countries' safe motherhood efforts is the presence of skilled birth attendants, a capable referral system and basic or comprehensive emergency obstetric services.

Progress in most other countries has been slow, and maternal mortality and morbidity remain tragically high in several regions, including in most of sub-Saharan Africa and the poorer parts of South Asia. While some gains in combating maternal death and illness are expected in the next 10 years, current interventions will need to be scaled up and more resources directed towards them if significant inroads are to be made to protect women's lives and health.

Causes and Consequences

COMMON CAUSES. WHO defines maternal mortality as "the death of a woman while pregnant or within 42 days of termination of pregnancy, irrespective of the duration or site of the pregnancy, from any cause related to or aggravated by the pregnancy or its management, but not from accidental or incidental causes."[12]

The causes of maternal death are remarkably consistent around the world.[13] Some 80 per cent are due to direct obstetric complications: haemorrhage, sepsis, complications of abortion, pre-eclampsia and eclampsia, and prolonged/obstructed labour. About 20 per cent of deaths have indirect causes, generally existing medical conditions that are aggravated by pregnancy or delivery. These include anaemia, malaria, hepatitis and, increasingly, AIDS.

VAST GAP IN IMPACTS. But huge differences—up to a hundred-fold—exist in the risk of pregnancy between women in rich and poor countries, the highest differential of any public health indicator monitored by WHO. The lifetime risk that a woman in West Africa will die in pregnancy or childbirth is 1 in 12. In developed regions, the comparable risk is 1 in 4,000.[14]

Within countries, poverty dramatically increases a woman's chances of dying during or soon after pregnancy.[15] Indeed, alarming gaps exist in many countries between wealthier and poorer women and safe motherhood care. In Bangladesh, Chad, Nepal and the Niger, elite populations have high rates of skilled attendance while for almost all other women giving birth with a skilled attendant is a rarity (national rates of skilled attendance in these countries are among the lowest in the world). In other countries where rates of skilled attendance are fairly high, including Brazil, Turkey and Viet Nam, the poorest women are still the least likely to receive such care.[16]

Because they receive prompt and effective treatment, women in the developed world rarely die or experience permanent disabilities from pregnancy-related problems.

THE "THREE DELAYS". In the process of home deliveries, experts have classified the underlying causes of maternal mortality according to the "three delays" model: delay in deciding to seek medical care; delay in reaching appropriate care; and delay in receiving care at health facilities.

The first delay stems from a failure to recognize danger signs. This is usually a consequence of the absence of skilled birth attendants, but it may also stem from reluctance within the family or community to send the woman to a care facility due to financial or cultural constraints.

The second delay is caused by a lack of access to a referral health facility, a lack of available transport or a lack of awareness of existing services. The third delay relates to problems in the referral facility (including inadequate equipment or a lack of trained personnel, emergency medicines or blood).

This is why maternal mortality reduction pro-grammes should give priority to the availability, accessibility and quality of obstetric facilities. All countries that have reduced maternal mortality have done it through a dramatic increase in hospital deliveries.

Maternal Morbidity

The same factors that result in maternal mortality make pregnancy- and childbirth-related illness and injury the second leading cause after HIV/AIDS of lost years of healthy life among women of reproductive age in developing countries—accounting for nearly 31 mil-lion "disability-adjusted life years" lost annually.[17]

While the incidence and prevalence of maternal morbidity are not well understood, an often-used estimate is that 15 per cent of pregnant women will experience complications of pregnancy or delivery serious enough to require emergency obstetric care in a health facility.[18]

Direct causes of maternal morbidity are obstetric complications during pregnancy, labour, or the post-partum period due to interventions, omissions or incorrect treatment. Indirect maternal morbidity results from previously existing conditions or disease, aggravated by pregnancy; this type of disability may occur at any time and continue throughout a woman's life. Maternal morbidity can also be psychological, most often manifested by depression, which can result from obstetric complications, interventions, cultural practices or coercion.[19]

Obstetric Fistula

Obstetric fistula, a devastating maternal morbidity, is one of the most neglected issues in international reproductive health. Caused by prolonged and obstructed labour, a fistula is a hole that forms between a woman's vagina and bladder or rectum, leaving her with chronic incontinence. In most cases, the baby dies.

Women with fistula are unable to stay dry. The smell of urine or faeces is constant and humiliating. The social consequences are equally tragic: many girls and women with fistula become ostracized by society, abandoned by their husbands and blamed for their condition.

Fistula is more than a women's health problem. It typically affects the most marginalized members of society—poor, illiterate girls and young women living in remote areas. Root causes include early childbear-ing, malnutrition and limited access to emergency obstetric care. Widespread in sub-Saharan Africa, South Asia and some Arab States, fistula affects more than 2 million girls and women worldwide, with an estimated 50,000 to 100,000 new cases annually.[20]

Fistula is preventable through efforts to postpone early marriage and pregnancy, increased access to

| 18 | **CAMPAIGN TO END FISTULA** |

In 2003, UNFPA launched a global Campaign to End Fistula with the aim of making fistula as rare in developing countries as it is in the industrialized world. The campaign raises awareness of the issue, conducts needs assessments, and expands services for prevention and treatment.

A key goal is to highlight the importance of emergency obstetric care for all pregnant women. Working with govern-ment partners, NGOs and the international community, the campaign is now under way in more than 25 countries.

In Bangladesh, where some 70,000 women are living with obstetric fistula, a Fistula Repair Centre at the Dhaka Medical College has been established. The centre will help manage cases and train service providers in South Asia.

In Zambia, the campaign is supporting the fistula repair unit at Monze Mission Hospital. Initial efforts have con-tributed to increased awareness of the problem and improved quality of care.

In Sudan, a national campaign was launched under the slogan "Fistula: We MUST Care." UNFPA has purchased medical equipment and supplies for the Fistula Centre in Khartoum, one of the country's only facilities.

family planning services, and timely access to emergency obstetric care—particularly Caesarean section to relieve obstructed labour. It is treatable through surgical repair, with success rates as high as 90 per cent for uncomplicated cases. Most women return to a full and normal life after counselling, social rehabilitation and reintegration in their community.

However, the cost of transportation to reach help, corrective surgery and recovery care (about $300) is often prohibitive. Treatment facilities are few and far between, and there are not enough specially trained doctors and nurses available.

INFANT MORTALITY. Where mothers are at high risk of maternal death or illness, their children are at risk, too. Neonatal and infant deaths can result from poor maternal health and inadequate care during pregnancy, delivery and the critical immediate post-partum period.[21] Infections and birth asphyxia and injuries account for the majority of neonatal mortality, but low birth weight, complications from delivery and congenital malformations also contribute.

Reducing Maternal Mortality and Morbidity

Every country that has achieved low maternal mortality has developed a systematic approach involving skilled attendants for routine delivery, emergency obstetric care to treat complications, and referral and transport systems that ensure access to life-saving care.

Upgrading medical facilities so that women will seek care and expanding services where they are overloaded are among the first steps in preventing maternal death and disability. Particular attention must be given to reaching the poor, populations isolated by location and those affected by war and natural disasters.

Mobilizing families, communities and nations to support women during pregnancy and childbirth, through strengthened policy, legislative and regulatory frameworks for maternal health, is also crucial.[22]

Family planning is also critical to reducing maternal mortality and morbidity. Satisfying the existing unmet need for contraception would reduce pregnancies worldwide, causing maternal mortality to drop by 25-35 per cent.[23] Reducing adolescent pregnancies would also have an important impact.

Strategies to reduce maternal mortality also need the support of broader efforts to address women's health, among them better nutrition for women and girls to build resistance and avoid anaemia, combating infectious diseases such as malaria, and averting violence. Reproductive health interventions promote the health and survival of infants and provide an important link between goals for child and maternal health.

Difficulties in Measurement

Maternal mortality is difficult to measure for both conceptual and practical reasons, making all estimates subject to some degree of uncertainty. In many settings, record keeping is poor, and women's deaths and their causes may go unreported by families and communities.[24]

In general, the methods used to estimate maternal deaths (using vital registration systems, household surveys, census data and reproductive age mortality studies)[25] provide neither the input needed to design and monitor prevention programmes nor the information needed to assess the availability, quantity and quality of life-saving health services.

Measuring maternal morbidity is also a challenge for many reasons: facility-based data (hospital case reviews and discharge surveys, for example) have inherent biases;[26] clinical monitoring of large populations of pregnant and post-partum women is impractical; self-reports do not provide reliable information; and stigma and fear often make women reluctant to discuss maternal health and complications.

Despite the difficulties, a number of countries have, since the ICPD, put in place measures to improve data collection and record keeping to monitor maternal death and illness. These include Angola, Argentina, Bolivia, Cambodia, Cuba, Morocco, Mozambique, Namibia, Nicaragua, the Philippines, Senegal, Sri Lanka and Zimbabwe. Saint Vincent and the Grenadines holds an annual perinatal morbidity and mortality conference to analyse national data.[27]

Holistic Responses

Dramatic reductions in maternal mortality in Sri Lanka (by half in three years) and Malaysia (by three fourths in 20 years) resulted from the phased develop-

19 MEASURING PROGRESS IN PROVIDING CARE

The Millennium Development Goal 5, improve maternal health, has a target of reducing the maternal mortality ratio by three quarters between 1990 and 2015. Besides the maternal mortality ratio—which does not address maternal morbidity—the indicator chosen to measure progress is the proportion of births attended by skilled health personnel; this does not address the possibility of a woman having a life-threatening complication requiring emergency care.

To measure these aspects, the Millennium Project Task Force on Maternal Health and Child Health has recommended as an additional target the goal of the ICPD: universal access to reproductive health services by 2015 through primary health care systems. The Task Force further suggests that additional UN process indicators be used, including the number of functioning comprehensive and basic emergency obstetric care facilities per 500,000 population and the proportion of births taking place in emergency facilities (at appropriate levels of care). They also urge attention to ensuring equitable access to these facilities.

ment of widespread, accessible networks of facilities able to treat obstetric emergencies, complemented by the training and appropriate deployment of professional midwives, with closely linked back-up emergency obstetric services.[28] Expenditures were not high, but attention was paid to developing a primary health care system that reached all parts of the population, regardless of ethnicity, class or urban-rural difference, in line with the ICPD Programme's recommendations.[29] In these efforts, both countries have tried to find the appropriate mix of private versus public expenditures.[30]

Haiti's Ministry of Health has established a Committee for Reduction of Maternal Mortality, which has developed a national plan. A model of comprehensive emergency obstetric care, developed in eight hospitals, includes infection prevention, post-abortion care and the integration of maternity services with family planning methods and counselling.[31]

Antenatal Care

Although an exclusive focus on care during pregnancy has not been shown to have a direct impact on maternal mortality, antenatal care provides an important entry point for women to the health care system. It presents an opportunity to assess the future mother's overall condition, diagnose and treat infections, screen for anaemia and HIV/AIDS, enrol women in programmes to prevent transmission of HIV to infants, and prevent low birth weight. Women who get antenatal care are also more likely to have a skilled attendant present during childbirth.

PROGRESS AND NEEDED SERVICE LINKS. Some progress has been made in expanding rates of antenatal coverage since the ICPD. According to a recent report from WHO and UNICEF, the number of pregnant women receiving antenatal care from a skilled health provider has grown 20 per cent since 1990. The greatest increase, 31 per cent, has been in Asia, while the number of women getting antenatal care in sub-Saharan Africa has grown just 4 per cent.

More than half of all women in the developing world now receive at least four antenatal visits during pregnancy (the number recommended by WHO), although those with less education are vastly under-represented. Women with secondary schooling are two to three times more likely to receive antenatal care as women with no education. Poor women, too, are far less likely to receive antenatal care, as with all health services.[32]

While good quality antenatal care can improve women's health in the period immediately before and after birth, it does not have a significant impact on maternal death risks unless it is linked with delivery care.[33] Many countries are tying expansions in antenatal care with other safe motherhood services. For example, Mongolia, with its great distances and harsh weather, has created 316 maternal rest homes where women herders can stay and receive essential care during the weeks before delivery.[34]

Skilled Attendance

The majority of maternal deaths are due to unexpected complications. But attendants with the skills to respond are present at only about half of deliveries worldwide. Skilled attendance for all births is the only way to ensure emergency obstetric care for all those with complications. Skilled attendance during labour, delivery and the early post-partum period could reduce an estimated 16 to 33 per cent of deaths due to obstructed labour, haemorrhage, sepsis and eclampsia.[35]

A skilled attendant is a professionally trained health worker—usually a doctor, midwife[36] or nurse—with the skills to manage a normal labour and delivery, recognize complications early on and perform any essential interventions, start treatment and supervise the referral of mother and baby to the next level of care if necessary. Trained and untrained traditional birth attendants (TBAs) are not considered skilled attendants.[37]

A skilled attendant can influence maternal mortality by utilizing safe and hygienic techniques during delivery. However, these measures will not prevent most life-threatening infections, which are due to delayed treatment of complications such as prolonged labour, ruptured uterus or retained products.[38]

BACK-UP SYSTEMS AND TRAINING. Skilled attendants are limited to a narrow range of interventions when deliveries take place in the home. To be effective, skilled attendance requires adequate supplies,

equipment and infrastructure as well as efficient and effective systems of communication and referral to emergency obstetric care facilities. Political support and appropriate policies—including pre- and in-service training, supervision and health system financing—are also critical.[39] There is a wide variation in how much skilled attendants are supported and supervised by health care systems.

A number of countries have taken steps since the ICPD to improve training of skilled attendants, and also to increase their numbers (and capacities) in rural and other underserved areas. In Iran, for example, rural midwives receive theoretical and practical training for six months and are required to have managed at least 20 deliveries before qualifying as midwives. In Panama, provision of training for midwives working in rural areas and with indigenous populations is a priority.[40]

Emergency Obstetric Care

Any woman can experience complications during pregnancy. But virtually all obstetric complications can be treated. Low maternal mortality ratios are due, in large part, to the fact that complications are identified early and are treated.

COMBATING THE "THREE DELAYS". The "three delays" model has proven useful in designing programmes to manage obstetric complications. Overcoming delays in deciding to seek care, in reaching appropriate care facilities and in receiving care at those facilities requires sequential procedures—from antenatal care and preparation to attended births with referral capabilities.

Health services related to emergency obstetric care are categorized as basic and comprehensive.[41] Basic emergency functions, performed in a health centre without an operating theatre, include: assisted vaginal delivery; manual removal of the placenta and retained products to prevent infection; and administering antibiotics to treat infection and drugs to prevent or treat bleeding, convulsions and high blood pressure.

Comprehensive services require an operating theatre and are usually provided in a district hospital. These include all the functions of a basic emergency facility, plus the ability to perform surgery (Caesarean section)

| 20 | **IMPROVING MATERNAL HEALTH IN RURAL SENEGAL** |

To get from the village of Goudiry to the regional hospital in Tambacoumba, Senegal, women in labour had to travel 70 kilometres along a rough dirt road, often in donkey carts. Eight out of ten with complicated pregnancies didn't get help in time, and many died.

That was before 2001, when with UNFPA support Goudiry's tiny health clinic was expanded into an obstetric care centre with the equipment and personnel to handle blood transfusions and Caesarean sections. Already, the model clinic has saved more than 100 women.

An anaesthetist, 17 nurses and several trained community workers offer outreach services, including information about reproductive health issues. They also deliver contraceptive supplies to the surrounding areas.

Senegal's maternal mortality ratio is nearly 700 deaths per 100,000 live births. It averages only one gynaecologist per 30,000 women of reproductive age, and most work in the capital. Rural women give birth to five or six children on average. Severe bleeding and eclampsia are the leading causes of maternal death. Early marriage, female genital cutting and sexually transmitted infections are additional factors that complicate childbirth for many women.

to manage obstructed labour and to provide safe blood transfusion to respond to haemorrhages.

A number of countries are seeking to increase the number of basic and emergency obstetric care facilities as well as to bolster the capacity of staff and the quality of care provided. For example, with UNFPA support, Guinea-Bissau assessed needs for emergency obstetric care and has made plans to increase the number of facilities offering basic emergency care and those offering comprehensive care.[42]

QUALITY SERVICES. The quality of emergency obstetric care is key to success. Services must be available 24 hours seven days a week, and have well-trained and motivated staff, essential supplies and logistics in place, functioning transport and communication systems and ongoing monitoring.

A number of countries have put priority on improving access to emergency obstetric care, and raising its quality. Lebanon and Oman have strengthened their referral services. El Salvador has developed quality obstetrical model services in hospitals and health units. In Jamaica, access to emergency obstetric care, including special facilities for transportation and referral to higher levels of care, is provided in each district.[43]

Since the ICPD, various countries in sub-Saharan Africa have introduced training for health staff in essential obstetric care. These include: Angola, Benin, Burundi, Cameroon, Chad, Côte d'Ivoire, Guinea, Kenya, Lesotho, Liberia, Mozambique, Namibia, the Niger, Senegal, Swaziland and Zambia.

In Morocco, providers have been trained to use new protocols for treating obstetric emergencies. Comprehensive services have been established in five rural hospitals, and ten provincial hospitals have improved the quality of emergency obstetric care, resulting in a significant increase in the number of women receiving appropriate care.[44]

IMPROVING TRANSPORT AND REDUCING OTHER BARRIERS. Poor families are often unable and sometimes reluctant to find or pay for transport to a medical facility when a woman goes into labour. The Mother Friendly Movement in Indonesia has helped communities recognize the need for and establish emergency transport systems for women in labour.[45]

New efforts seek to understand obstacles to and promote the use of available care. A partnership of Canadian and Ugandan medical associations, undertaken as part of the Save the Mothers Initiative of the International Federation of Gynaecology and Obstetrics (FIGO), worked in the rural district of Kiboga in Uganda to improve emergency obstetric care and its use.

The number and capabilities of skilled attendants in the district hospital were increased, and local dispensaries made care available 24 hours a day; health facilities were upgraded and stocked; and workshops were held to improve health workers' attitudes towards community members. UNFPA provided two ambulances.

As a result of the interventions, met need for treatment of women with obstetric complications rose from 4 per cent in 1998, when the project began, to 47 per cent in 2000. Maternal deaths dropped from 9.4 per cent of those receiving emergency obstetric care in 1998 to around 2 per cent in 1999 and 2000.[46]

Post-abortion Care

About 19 million of the estimated 45 million induced abortions performed annually are unsafe (done by untrained people in less-than-hygienic circumstances); nearly 70,000 women die as a result, representing 13 per cent of pregnancy-related deaths.[47]

Many national health services dedicate a high percentage of beds in second- and third-level facilities to accommodate the large number of women who require post-abortion emergency treatment. In sub-Saharan Africa, up to 50 per cent of gynaecological beds are occupied by patients with abortion complications.[48]

The ICPD, in a groundbreaking consensus, called for all women to have access to treatment for abortion-related complications, post-abortion counselling, education and family planning services, regardless of the legal status of abortion.[49]

Post-abortion care is cost-effective, reduces repeat abortions and helps individuals meet their reproductive intentions.[50] Many countries now recognize the contribution it can make to saving women's lives. For example, Kenya's 1997 reproductive health service guidelines state, "The prompt treatment of post-abortion complications is an important part of health care that should be available at every district-level hospital." [51]

An international consortium has adopted a post-abortion care model, which aims to help women avoid further unwanted pregnancies and other reproductive health problems in addition to providing for their emergency needs. First developed by the NGO Ipas, the model includes: emergency treatment for complications of miscarriage or induced abortion; family planning counselling and services; management of sexually transmitted infections; counselling tailored to each woman's emotional and physical needs; and community and service provider partnerships. [52]

At least 40 countries have initiated post-abortion care programmes since the ICPD, including Honduras, Malawi, Mexico and Zimbabwe. [53] Facilities offering post-abortion care in Egypt increased significantly between 1999 and 2001, [54] and treatment of complications of incomplete abortion is now part of essential obstetric care protocols. [55]

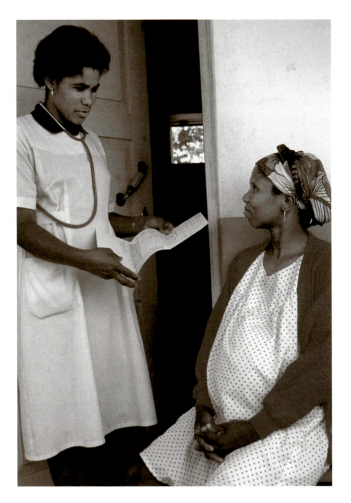

Myanmar's Department of Health integrated post-abortion care and contraceptive services into existing health care at the township level. Midwives now make follow-up home visits to women with abortion-related complications and provide family planning methods when requested. [56]

SERVICES FALL SHORT. Yet, lack of access to treatment for incomplete abortion remains a major problem. A study in Ethiopia found that only 16 of 120 health centres were able to respond with emergency transport to assist a woman needing post-abortion care. [57] Documenting the need for post-abortion care is difficult. One hospital study in Kenya found abortion was an important cause of admission but was rarely recorded as the cause in the death register, "a fact likely due to the stigma attached to abortion-related mortality". [58]

The social taboos surrounding abortion [59] and the penalties for both women who seek abortions and those who provide them are further challenges in many countries, even where post-abortion care is legal. A study in Zimbabwe found that the most common reason given for not seeking prompt medical attention for abortion complications was fear of being reported to the police. [60]

Adolescents undergo a major share of illegal abortions. For them, stigma, shame and disapproval from providers can be intense and may discourage many from seeking treatment.

Quality of Maternal Health Care

As with family planning, quality is also important in maternal health programmes, [61] and can increase the likelihood that women facing obstetric emergencies will go to health facilities for life-saving care. A study of 164 households in Mexico where a maternal death occurred found that perceived quality of health care was a significant factor in a woman in labour delaying seeking medical care. [62]

Other studies found similar concerns among potential users. In Bolivia, women say they consider respectful treatment to be crucial and that condescending provider attitudes are the greatest deterrent to the use of maternal health services. [63] A study in Yemen of randomly selected households found that both rural and urban women preferred to deliver at home, despite acknowledging the importance of their medical needs, because they feared bad experiences or had prior bad experiences with institutional deliveries.[64]

Since the ICPD, attention to quality services has increased significantly, and many countries have launched initiatives to improve maternal health care. For example, with donor assistance, Azerbaijan initiated a Safe Motherhood and Newborn Care project, which includes capacity-building of health staff, raising the awareness of women and adolescent girls regarding healthy lifestyles, promoting a mother- and baby-friendly environment at maternity units, and giving special attention to vulnerable populations. Several governments in the Caribbean have established maternal and child health committees or technical advisory groups, whose members include doctors, nurses and social workers.[65]

Training providers is an important element in providing improved care and overcoming clients' reservations about seeking out available services. It can also have positive impacts on maternal health outcomes.[66] For example, a training programme in Moldova led to sites welcoming fathers or family members as support persons for women in labour, maternity hospitals allowing family visits post-partum, women reporting a decline in invasive practices, and more parents attending new childbirth classes.[67]

And after 24 trainers were trained in the Ukraine, induction of labour and Caesarean sections were less frequent, and providers reported that they felt that the women giving birth were happier.[68] Training providers

in Russia resulted in a huge increase in clients spending the night in a hospital following delivery, from 0.5 per cent to 86 per cent one year later.[69]

Men and Maternal Health

Men's social support for women during pregnancy, labour and delivery results in positive maternal health outcomes.[70] In Guatemala, husbands often provide care and support during pregnancy, accompany their wives on antenatal visits, and are generally present or nearby during births at home; while they may not be with their wives who deliver in hospitals, they take care of baby-related preparations.[71] Research in Egypt shows that involving husbands in post-abortion care speeds their wives' recovery and leads to greater contraceptive use afterward.[72]

Many men who want to be involved in the health of the women in their lives, however, are impeded by hospital regulations, work schedules or poor communication with their partners.

Getting male partners, extended families and community members to give more priority to women's health care during pregnancy and the post-partum period has been a successful tactic in many programmes. For instance, the Pati Sampark project in India gives husbands information about pregnancy and birth, and points out specific roles that they can fill, including providing household help during pregnancy and making plans involving transportation.[73]

UNFPA and Safe Motherhood

To reduce maternal deaths and ill-health, UNFPA supports a range of activities in developing countries,[74] taking into consideration poverty reduction, governance, economic and social development and health system reform.

NEEDS ASSESSMENTS. Working with the Averting Maternal Death and Disability (AMDD) programme managed by Columbia University's Mailman School of Public Health and supported by the Bill & Melinda Gates Foundation, UNFPA has conducted needs assessments in Cameroon, Côte d'Ivoire, India, Mauritania, Morocco, Mozambique, Nicaragua, the Niger and Senegal, among others.

| 21 | **MEETING CLIENTS' NEEDS IN PERU** |

At a MaxSalud primary health clinic in a peri-urban community in Peru, health managers were concerned by low utilization rates in 2000 among their mostly female clientele. An assessment found that women were reluctant to get reproductive health services from the male gynaecologist after the midwife transferred to another clinic. Prices were reduced, an aggressive publicity campaign was launched to advertise the new fees, and a certified midwife was reinstated as part of the clinic's quality improvement intervention. Service utilization rates rose dramatically. Female clients reported high satisfaction with women's health services and a preference for female providers.

In Nicaragua, the assessment of 125 facilities led to a range of improvements in three health regions: physical upgrades, publication of standards and protocols for care, staff training and efforts to improve service quality. Between 2000 and 2003, the proportion of women with complications receiving emergency care in these regions rose by one third.[75]

TRAINING. Training of doctors, nurses, midwives and anaesthetists in emergency obstetric care and post-abortion care is ongoing in all regions, along with training of service providers in record keeping and data collection. Health personnel have been trained to evaluate maternal deaths and complications. The Fund has also developed technical materials, a distance learning course and an Emergency Obstetric Care Checklist for planners.

INFRASTRUCTURE DEVELOPMENT. UNFPA has upgraded facilities and built new ones, provided equipment and supplies, and procured ambulances and radio communications.

In Rajasthan, India, with support from the UNFPA/AMDD initiative, 83 emergency obstetric care facilities were upgraded, covering a population of approximately 13 million people. Fifty-nine teams of health professionals were trained to provide emergency obstetric care; 12 were trained in infection prevention. Management information systems were introduced to improve monitoring and evaluation of services. In 2003, local television stations and newspapers started airing programmes related to safe motherhood. As a result of these efforts, the number of women treated for obstetric complications has increased 50 per cent in four years, and India has introduced similar interventions in other states.[76]

POLICY AND ADVOCACY. UNFPA, along with UNICEF and WHO, organized Vision 2010 to spotlight maternal and neonatal mortality in Central and West Africa, and initiated the Forum for Maternal Mortality reduction in Latin America with the Pan-American Health Organization. The Fund has worked with health ministries to develop emergency obstetric care guidelines and protocols, national safe motherhood policies and routine monitoring of referral systems.

COMMUNITY MOBILIZATION. The Fund trains volunteers and health promoters to educate communities about safe motherhood services and family planning and encourages community-sponsored transportation schemes.

8 Preventing HIV/AIDS

Over two decades into the AIDS pandemic, some 38 million people are living with HIV/AIDS and over 20 million people have died.[1] Despite expanding prevention activities, some 5 million new infections are occurring each year. In the hardest-hit countries, the pandemic is reversing decades of development gains.

In 1994, the ICPD Programme of Action noted the severity of the pandemic and projected that the number of people infected with HIV would "rise to between 30 million and 40 million by the end of the decade if effective prevention strategies are not pursued".[2]

The ICPD called for a multisectoral approach to AIDS that included raising awareness about the disastrous consequences of the disease, providing information on means of prevention, and addressing the "social, economic, gender and racial inequalities that increase vulnerability".[3] It recognized the harm of stigma and discrimination and the need to protect the human rights of people living with HIV/AIDS.

The Programme of Action also noted that the "social and economic disadvantages that women face make them especially vulnerable to sexually transmitted infections, including HIV".[4] In Africa today, women are 1.3 times more likely than men to be infected with HIV. Young women aged 15-24 are two and a half times more likely to be infected than young men.

Reproductive health programmes were recognized as essential to preventing HIV through: prevention, detection and treatment of sexually transmitted infections; provision of information, education and counselling for responsible sexual behaviour; and ensuring a reliable supply of condoms.

In its five-year review of ICPD implementation in 1999, the United Nations took note of the worsening pandemic and called for increased resources and stepped-up efforts to combat its spread. Targets were set for reducing HIV prevalence among young people, and for expanding their access to information and services for preventing infection.

Given that more than three fourths of HIV cases are transmitted sexually and an additional 10 per cent are transmitted from mothers to children during labour or delivery or through breastfeeding,[5] linking HIV and reproductive health services is crucial. The pandemic has highlighted the urgent need to improve both primary health services and sexual and reproductive health services.

INTEGRATE HIV/AIDS PREVENTION WITH REPRODUCTIVE HEALTH SERVICES

Governments should ensure that prevention of and services for sexually transmitted diseases and HIV/AIDS are an integral component of reproductive and sexual health programmes at the primary health-care level.

—Key Actions for the Further Implementation of the Programme of Action of the International Conference on Population and Development, para. 68

Impact and Response

HIV/AIDS is taking a terrible toll on individuals and communities in countries with high prevalence. In some sub-Saharan African countries, one quarter of the workforce is infected with HIV. By one estimate, if 15 per cent of a country's population is HIV positive (a level nine countries are expected to reach by 2010), gross domestic product declines by about 1 per cent each year.[6] Using this measure, South Africa's GDP may fall by 17 per cent by 2010.

A recent report from the World Bank and Heidelberg University warns that the long-term impact of AIDS may be even more damaging.[7] AIDS destroys human capital by killing people in the prime of their lives and also affects the way knowledge and skills are transferred from generation to generation. Furthermore, premature adult mortality associated with AIDS weakens investments in education and reduces the proportion of families that can afford to send their children to school.

Fewer than one in five people at high risk of HIV infection have access to proven prevention interventions, according to a 2003 report by the Global HIV Prevention Working Group, an international panel of AIDS experts. Dramatically scaling up proven prevention strategies could avert 29 million of the 45 million new HIV infections expected by 2010, the report said.[8]

Treatment regimes for HIV improved throughout the 1990s, but their cost remained prohibitive for all but the wealthiest countries. While there is now a concerted effort to expand access to treatment—including the WHO-led UNAIDS "3 by 5 Initiative" to reach 3 million people by 2005 and lower drug costs—the vast majority of infected people still do not have

---

23 | CONFRONTING INEQUALITY

To address the disproportionate impact of HIV/AIDS on women and girls, UNAIDS launched the Global Coalition on Women and AIDS at a February 2004 meeting chaired by UNFPA Executive Director Thoraya Obaid. The advocacy initiative will focus on preventing new HIV infections among women and girls, promoting equal access to HIV care and treatment, accelerating microbicides research, protecting women's property and inheritance rights and reducing violence against women.

UNAIDS, the United Nations Development Fund for Women (UNIFEM) and UNFPA issued a joint report in July 2004, *Women and AIDS, Confronting the Crisis*. It calls on governments and the world community to:

- **Ensure that adolescent girls and women have the knowledge and means to prevent HIV infection** through advocacy campaigns that convey basic facts about women's heightened physiological vulnerability and dispel harmful myths and stereotypical notions of masculinity and femininity, warn that marriage does not necessarily offer protection from HIV transmission, and involve both young men and women in promoting sexual and reproductive health.

- **Promote equal and universal access to treatment** by ensuring that women make up 50 per cent of people able to access expanded treatment interventions, increasing access to confidential voluntary counselling and testing (VCT) services that take into account unequal power relations and encourage partner testing, expanding reproductive and sexual health services, and training health providers in gender-sensitive care and treatment.

- **Promote girls' primary and secondary education and women's literacy** by eliminating school fees, promoting zero tolerance for gender-based violence and sexual harassment, offering literacy classes for women that focus on HIV/AIDS and gender equality, providing life skills education both in and out of school, and creating curricula that challenge gender stereotypes and promote girls' self-esteem.

- **Relieve the unequal domestic workload and caring responsibilities of women and girls for sick family members and orphans** by providing social protection mechanisms and support for caregivers, promoting more equitable gender roles in the household, distributing home-care kits, and establishing community fields and kitchens to supplement individual household responsibilities.

- **End all forms of violence against women and girls** by undertaking media campaigns on zero tolerance for violence, male responsibility and respect for women, and dangerous behaviour norms, and by providing counselling and post-exposure prophylaxis to all who experience sexual violence.

- **Promote and protect the human rights of women and girls** by enacting, strengthening and enforcing laws protecting their rights, reporting violations to the UN Committee on the Elimination of Discrimination Against Women, protecting women's property and inheritance rights, and supporting free or affordable legal services for women affected by HIV/AIDS.

access to antiretroviral (ARV) therapy, which can transform AIDS into a chronic disease.

A June 2004 report[9] by the Global HIV Prevention Working Group stressed the importance of integrating HIV prevention interventions into expanding treatment programmes. Increased availability of ARVs, the report stated, will bring more people into health care facilities where they can be reached by HIV prevention messages. But it could also lead to an increase in risky behaviour unless prevention counselling is incorporated into treatment programmes. The group recommended making VCT available in all health care settings where people have access to ARVs.

FEMINIZATION OF THE PANDEMIC. Half of all adults living with HIV/AIDS are female, compared to 41 per cent in 1997. In sub-Saharan Africa, the most affected region, the figure is nearly 60 per cent. The rising rates of infection among women and adolescent girls reflect their greater vulnerability, due to both biological and social factors. Gender inequities and male domination in relationships can increase women's risk of infection and limit their ability to negotiate condom use. Poverty leads many women and girls into unsafe sexual relations, often with older partners.

This "feminization" of the epidemic is further exacerbated by women's roles as managers of the household and primary caregivers for family members infected with HIV. Other factors that make the impacts disproportionate include the legal, economic and social inequalities women often face in the areas of education, health care, livelihood opportunities, legal protection and decision-making.

COUNTRIES RESPOND. Three fourths of the countries responding to UNFPA's 2003 global survey reported adopting a national strategy on HIV/AIDS and 36 per cent said they had specific strategies aimed at high-risk groups. Many countries have established national AIDS commissions and developed policies and programmes to address the impact of the pandemic. A growing number of countries are taking a multisectoral approach, involving a wide range of ministries and increased involvement of NGOs. But just 16 per cent reported having passed legislation in support of HIV/AIDS efforts.

Linking HIV Prevention and Reproductive Health Programmes

Since a majority of HIV transmission takes place through sexual contact, reproductive and sexual health information and services provide a critically important entry point to HIV/AIDS prevention. They also provide a conduit to and delivery point for programmes of care and treatment.

Reproductive health services can help prevent HIV transmission by: providing education on risks to influence sexual behaviour; detecting and managing sexually transmitted infections (STIs); promoting the correct and consistent use of condoms; and helping to prevent mother-to-child transmission.[10] Linking HIV prevention and the prevention and treatment of STIs with family planning and maternal health interventions can improve outreach, reduce stigma and save money by using existing resources and infrastructure.

A number of initiatives, primarily in Africa, have sought to link HIV prevention and reproductive health programmes.[11] But in many settings, the two programmes are not linked to each other.

CALL TO COMMITMENT. In June 2004, UNFPA, UNAIDS and Family Care International convened a high-level global consultation, involving health ministers, parliamentarians, ambassadors, leaders of UN agencies, donor organizations, community and NGO leaders, young people and people living with HIV. The meeting resulted in a Call to Commitment that emphasized "the urgent need for much stronger links between sexual and reproductive health and HIV/AIDS policies, programmes and services".

Closer links—including provision of reproductive health information and services to all people reached by HIV/AIDS programmes, and of HIV/AIDS information and services to all people reached by reproductive health programmes—are critical to success in both areas and to achieving the MDGs, the call stated, and "will result in more relevant and cost-effective programmes with greater impact".

EDUCATION ON RISKS. Up to now, few family planning programmes have focused on enabling providers to deal with sexuality issues. A recent study in the United Republic of Tanzania, for example, found that

HIV/AIDS was mentioned to family planning clients only briefly during counselling for informed choice, when "women were told that condoms prevent STIs such as HIV, and other methods do not".[12]

The International Planned Parenthood Federation has trained a number of family planning associations on sexuality, gender and quality of care.[13] These experiences have shown that counselling on sexuality can take place if providers have adequate training. In studies in Kenya and Zambia, however, most providers said they doubted their ability to adequately counsel clients on contraceptive needs in light of HIV.[14]

Programmes need to train all providers to help clients assess HIV risk and counsel them about avoiding both disease and unintended pregnancy. Training must also include correct information on contraceptive methods to help dispel the myths and rumours that abound.

MANAGING STIs. The presence of one or more STIs significantly increases the risk of becoming infected with HIV. A recent U.S. study found that treating an STI in an HIV-infected person can result in a 27

per cent reduction in HIV transmission, without any other behaviour change.[15] A study in the United Republic of Tanzania in the mid-1990s showed that preventing and treating STIs could prevent about 40 per cent of new HIV infections.[16] The study also showed that community-based publicity, partner notification and treatment efforts could reduce the spread of STIs.

Reproductive health programmes can educate service users about STIs, their symptoms and transmission, and healthy behaviour. Detecting and managing STIs has proven difficult, however. Most women with STIs have no symptoms, and as a result, efforts to identify and treat them in reproductive health settings have proven to be of little benefit for women.[17] Consequently, STIs have not received adequate attention in either reproductive health or HIV programmes,[18] and health ministries have done little to ensure that they are included.[19] The recent development of cheaper tests for common STIs could help remedy this deficiency.

Condoms

Both male and female condoms are key components of prevention efforts among the sexually active. Programmes can encourage the use of condoms to protect against unintended pregnancy and other STIs as well as HIV, and inform clients that non-barrier contraceptives do not prevent HIV transmission. Correct and consistent use are critical: in a Uganda study, none of the 350 women who reported consistent condom use became HIV positive, but incidence was significant among women who reported inconsistent use.[20]

But the challenges to wider condom use are significant. There are massive shortfalls in supply compared to current

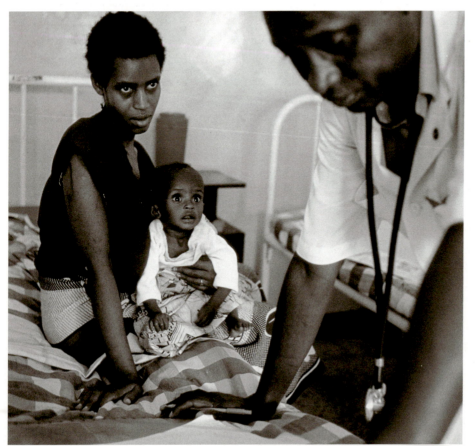

needs, frequent stock-outs, and limited resources for programming to instil safer sexual behaviours. Pervasive myths, misperceptions and fears about condoms also inhibit their use.

Ensuring a steady, affordable supply of high-quality condoms involves forecasting needs, procurement, logistics management and quality assurance. Promotion of condom use also requires an understanding of people's needs and preferences, and of the sociocultural environment of communities and countries.

THE FEMALE CONDOM. Introduced in the 1990s, the female condom has added to women's limited choice of means of protection. Wider use has been hampered by its price, which is 10 times higher than a male condom.[21] While evidence suggests that the female condom can be reused up to seven times if properly washed with disinfectant, WHO currently recommends one-time use until more data are available on the safety of reuse.[22]

More than 19 million female condoms have been supplied in more than 70 countries. Brazil, Ghana, Namibia, South Africa, Zambia and Zimbabwe all have large programmes. A study in Costa Rica, Indonesia, Mexico and Senegal found that the female condom is most acceptable where men already support family planning and perceive that their peers would support use of the method, where sex workers already have skills in negotiating safer sex, and where the female condom is considered preferable to the male condom.[23] This suggests that "marketing should focus on women who can successfully negotiate female condom use with their partners, as well as encouraging men to accept the device."[24]

Research in this area continues, and there are indications that a new generation of less-expensive female condoms could be available within a few years.

DIFFICULTIES NEGOTIATING CONDOM USE. Data from the United Republic of Tanzania show that HIV/AIDS prevalence is greater among married, monogamous young women than among sexually active unmarried women. Married women, particularly young women, often cannot negotiate condom use even if they know their husband now has or previously had multiple

partners; they are more able to do so if condom use is widely accepted as an important component of overall protection.

Research in South Africa found that women who used a condom the last time they had sex were almost twice as likely as women who did not use one to know that condoms prevent HIV.[25] Women most likely to use condoms were also young, more educated and lived in urban areas. Women with casual partners were four times more likely and women with regular, non-marital partners were twice as likely to have used a condom than women who last had sex with their husbands.

Family planning and antenatal care providers need to assist clients in developing communication and negotiation skills to use with their partners so that they may adequately protect themselves from HIV transmission.

DUAL PROTECTION. Providers should be familiar with dual protection, a strategy for preventing both transmission of HIV or STIs and unintended pregnancy, through the use of condoms alone, the use of condoms combined with other methods for extra protection against pregnancy (dual method use), or the avoidance of sexual activities considered high-risk. The dual protection message can be expanded to include the safeguarding of fertility through STI prevention and control, a particularly relevant message for young women who have not begun childbearing.[26]

MICROBICIDES. Microbicides to protect women against HIV transmission are under development but are several years off. For those who lack the ability to negotiate condom use, introduction of a microbicide would greatly increase their ability to protect themselves against infection. The Global Campaign for Microbicides estimates that a product could be available as early as 2007 and that 2.5 million HIV infections could be averted over three years if a microbicide that is 60 per cent effective were used in developing countries.[27]

Voluntary Counselling and Testing

VCT programmes confidentially and sensitively let women and men know their status and risks, and promote healthy adaptations of behaviour. They are an effective means of preventing HIV transmission and an important entry point for treatment of HIV-related illnesses, prevention of mother-to-child transmission, tuberculosis control, and psychosocial and legal support.

Pilot projects in Côte d'Ivoire and India indicate that integrating VCT into sexual and reproductive health services reduces the stigma associated with HIV/AIDS, strengthens awareness of healthy sexual behaviour, and increases access to and utilization of services. Yet, all too often, VCT has been introduced in isolation from other services.[28]

PREVENTING MOTHER-TO-CHILD TRANSMISSION.
Reproductive health programmes can help prevent mother-to-child transmission of HIV by preventing infection among women, by ensuring that HIV-positive women and men have information on their options and risks so they can make informed choices, and by providing access to ARVs and to a range of contraceptive methods to help support those decisions.

In many settings, pregnancy is often one of the few times when women access health services, providing an excellent opportunity for HIV prevention, especially through counselling and VCT. Integrated services for HIV prevention and maternal health can promote condom use, manage STIs, and provide prenatal and post-delivery care, safe delivery and counselling on infant feeding.

There is a great need to scale up initiatives for the prevention of mother-to child-transmission, which currently reach only a small percentage of women.[29]

Key Challenges

EXPANDING ACCESS TO TREATMENT. In the past two years, pledges of expanded funding for treatment efforts and the increased availability of generic ARV drugs have changed the focus of HIV/AIDS programmes. WHO and its partners in the Joint United Nations Programme on HIV/AIDS (UNAIDS) have undertaken the enormous challenge of treating 3 million people with ARVs by 2005.[30] In Africa in 2003, only 100,000 people were receiving ARVs, just 2 per cent of an estimated 4.4 million needing treatment. In South and East Asia, 7 per cent of those who need treatment were covered by ARV services.

The United States has pledged to provide funds to 15 African and Caribbean countries to treat 2 million people over the next five years, as part of a new initiative intended to prevent 7 million new infections and to provide care and support for 10 million people living with HIV/AIDS.

WHO and UNAIDS recognize that treatment initiatives are a long-term commitment. "Lifelong provision of therapy must be guaranteed to everyone who has started antiretroviral therapy. Thus, 3 by 5 is just the beginning of ongoing antiretroviral therapy scale-up and strengthening of health systems."[31]

Rapid expansion of treatment will require immediate strengthening of health systems—which can benefit sexual and reproductive health programmes as well—and increasing the number of health care providers (a goal of the 3 by 5 Initiative).

Wider availability of treatment will contribute to HIV prevention efforts, as those on ARVs are less likely to spread the virus, and their provision creates opportunities to communicate prevention messages and provide condoms. But it is imperative to increase support for prevention programmes as well, and to integrate prevention into treatment initiatives.

Another key need is to strengthen universal precautions (including safety procedures and proper disposal of gloves and sharp objects). Currently, even when service providers know the precautions to take against HIV transmission, they often lack protective gloves and ARVs (in case of needle sticks or other accidents). A recent study in Zambia found that clinics had no safety guidelines or post-exposure prophylaxis kits. When gloves were in short supply, clients were told to buy them. Addressing occupational risk may boost staff moral and improve client care.[32]

PREVENTION PRIORITIES. A recent study of incidence of HIV in Cambodia, Honduras, Indonesia, Kenya and Russia suggests that the focus of prevention activities should be based on a careful analysis of where infections are occurring and not simply on broad categories of at-risk groups.[33] For example, in Cambodia, new infections acquired as a result of commercial sex have fallen, while the proportion of new infections acquired in marriage has increased from 11 per cent to 46 per cent. Yet the initial transmission of HIV is still heavily linked to sex work and most people becoming infected are those whose partners had

---

26   REDUCING HIV RATES: LESSONS FROM UGANDA

HIV prevalence in Uganda peaked at around 15 per cent in 1991 and then fell to 5 per cent by 2001. This decline has been attributed to prevention efforts on several fronts, in a combined public health approach to behaviour change described as "ABC"—promotion of abstinence, partner reduction (also called "be faithful"), and consistent and correct condom use, coupled with VCT services.

Abstinence. Data indicate that young Ugandans are increasingly starting sexual relations at a later age. Between 1989 and 2000, the age at onset of sexual relations increased from 15.9 to 16.6 years among young women and from 17.3 to 18.5 among young men. Educational efforts promoting abstinence and increased knowledge about the disease both played a part.

However, the proportion of people who were sexually active dropped substantially only among adolescent women aged 15-17 and not other groups. About half of all unmarried women were sexually experienced in 2000, the same proportion as in the late 1980s. Just over half of unmarried men were sexually experienced, a one third decline.

Be faithful (reduce the number of partners). Monogamy increased between 1989 and 1995, especially among sexually active unmarried women, but changed little thereafter. The proportion of women with multiple partners was cut in half, but from already low levels. In contrast, more than 25 per cent of active unmarried men had multiple partners. Married men with multiple partners increased in some age groups but decreased in others.

Condom use. Dramatic changes occurred in condom use, particularly among unmarried men and women, in the 1990s. Before 1989 use by either sex was negligible. By 1995 it increased to 8 per cent among women and 11 per cent among men. Later increases have been striking, particularly among the youngest (and most vulnerable) age groups.

Condom use by sexually active women aged 15-17 increased from 6 per cent to 25 per cent and by those 18-19 from 3 per cent to 12 per cent. For men aged 15-17, condom use rose from 16 per cent to 55 per cent and in those 18-19, from 20 per cent to 33 per cent. Recent data suggest continued increases in condom use.

high-risk behaviours in the past. Thus, prevention needs to focus on both sex work and prevention within marriage.

FAMILY PLANNING. It is also critical that an increased emphasis on HIV/AIDS prevention and treatment does not come at the expense of other sexual and reproductive health information and services. Data from the 2003 Demographic and Health Survey suggests this has occurred in Kenya, where the family planning programme made gains in the 1990s. Contraceptive prevalence rose steadily from 27 per cent in 1989 to 39 per cent in 1998, but has not increased since.

In the context of high HIV prevalence, it is critical to continue support for family planning, which is a key component of reducing mother-to-child transmission. Recent studies in Kenya and Zambia found that family planning providers, clients for antenatal care and family planning, and HIV-positive women all saw an increased need for family planning to avoid unintended pregnancies.[34]

At the same time, providers need to respect the rights of all people, including those infected, to make their own decisions about having children and to have access to accurate information and humane treatment in order to do so. In many cases, HIV-positive women are told that they should not have children. This discriminatory treatment leads many such women not to disclose their status to health workers.

INTEGRATION CHALLENGES. A recent study found that "many health sector reforms have separated sexuality education, [reproductive health services] and STI/HIV/AIDS programmes from each other, making different ministries or segments of health ministries responsible for them, which also creates potential rivalry for budgetary control and funding."[35]

An assessment in the Kaolack region of Senegal in 2001 found little evidence of integration of family planning or maternal and child health care with STI/HIV/AIDS services. "The obvious lack of availability of HIV/AIDS services in the districts' health care and community structures attests once again that the lack of decentralization for these activities is

---

27 | **POSITIVE WOMEN: VOICES AND CHOICES**

The Positive Women: Voices and Choices advocacy-research project developed by the International Community of Women Living with HIV/AIDS is exploring the impact of HIV/AIDS on women's sexual and reproductive lives, challenging the violation of their rights and advocating improvements in policies and services.

The effort in Zimbabwe, one of three project countries, was carried out from 1998 to 2001. HIV-positive women were generally unaware of their risk before they were tested. Gender norms and economic dependence on their husbands or partners restricted the women's control over their sexual and reproductive lives. In the face of prejudices about HIV-positive women being sexually active and having children, they did not tell health workers about their status, making it difficult to address their needs. Using condoms in marriage was not considered appropriate.

The younger women wanted to have children, while older women with several children wanted to limit childbearing after their HIV diagnosis. Condom use and contraceptive use increased markedly among women who attended support groups. The project affirmed that HIV-positive women need better economic opportunities, pregnancy and delivery care integrated with family planning and STI/HIV-related services.

---

hindering any possibility of discussing integration on these levels."[36]

Reforms intended to strengthen health systems, including the creation of minimum services packages, should ensure that HIV/AIDS prevention and treatment services are included along with family planning.[37]

Health providers, who face mounting work loads and often staff reductions, must be enabled not only to provide contraceptives to clients, but also to identify related reproductive health problems. They need space to examine and counsel patients in private, and need supplies and equipment that are often missing from clinics in developing countries including gloves, speculums, spotlights and syringes for contraceptive injections.

Health providers also often need to be educated about HIV/AIDS to overcome bias, and taught to communicate prevention messages and to help individuals accessing services assess their risk of infection.

CHANGING BEHAVIOUR. Combating HIV/AIDS calls for addressing the underlying socio-economic, cultural and behavioural factors that contribute to its spread—including the lack of paid jobs, particularly for women, employment and migration policies that force many people to migrate for jobs, gender-based violence and trafficking of women.[38]

Behaviour change is gradual, multifaceted and needs to improve the health and reduce the risks of diverse married and unmarried young populations. As experience in Senegal and Uganda[39] shows, promoting responsible, voluntary and safe behaviour requires comprehensive and multisectoral efforts that foster partnerships involving central and local governments, the private sector, development partners, cultural leaders and a wide range of civil society organizations.

A project in Zimbabwe is seeking to offer a financial shield for girls to resist sexual liaisons with older men, often called "sugar daddies", who provide support in exchange for sex.[40] The programme offers vocational and life skills training, loans and jobs, linked to education about reproductive health and condom negotiation. A study will assess the impact of the programme on HIV, STIs, the onset of sexual activity and pregnancy.

REACHING HIGH-RISK GROUPS. Attention to reaching high-risk groups with HIV/AIDS prevention information, or treatment and care, has increased substantially in recent years. Most of the countries polled in the UNFPA global survey reported undertaking programmes to reach groups such as sex workers, injecting drug users, long-distance truck drivers, men who have relations with men, street children, soldiers and migrant workers, as well as adolescents and youth.

NGOs are often key partners or initiators. In Kenya, for example, a project that offers education and counselling on responsible sexual behaviour and condom use and provides economic alternatives has reached 15,000 sex workers and their clients. In Bangladesh, a variety of organizations provide sexual health services to commercial sex workers.[41]

Countries in Latin America have been particularly active in educating members of the armed forces to stem HIV/AIDS infection and further transmission.[42]

---

28 | **COMPREHENSIVE APPROACH TO HIV/AIDS PREVENTION IN SIERRA LEONE**

Sierra Leone recently emerged from more than ten years of war that seriously disrupted all sectors of society. Almost two thirds of the population was uprooted, rape and sexual abuse were widespread, and thousands of girls and women who lost their families have turned to sex work to survive. People's heightened vulnerability to HIV/AIDS has now become a priority concern.

UNFPA has responded through a coordinated initiative that targets different groups and involves a range of national actors and government offices. The overall aim is to reduce the risk of HIV/AIDS and other STIs, reduce women's need for commercial sex work, alleviate poverty and enhance family life and community security.

Conducted in partnership with the Government, UNAIDS, the UN Department of Peacekeeping Operations, UNIFEM and other partners, UNFPA's comprehensive approach includes the following:

- Workshops and other initiatives to promote HIV/AIDS prevention among the Sierra Leone police force, the Sierra Leone military, international peacekeepers, and demobilized soldiers;

- Promoting HIV/AIDS prevention among sex workers, through both health education and skills training to help them find other sources of income;

- Strengthening the capacity of partner NGOs;

- Promoting HIV/AIDS prevention among refugee and IDP populations;

- Creation of safe blood supply.

In 2003, the UN Security Council recognized the multi-party, multi-pronged initiative as a potential breakthrough in responding to HIV/AIDS in post-war reconstruction, bringing everyone—including ex-combatants, military personnel and international peacekeepers—together to promote improved health and HIV prevention. UNFPA and partners are applying similar approaches in neighbouring Liberia and in the Democratic Republic of the Congo.

COMBATING STIGMA AND BIAS. People living with HIV/AIDS still face stigma and legal and social discrimination in all regions. Ghana, South Africa and Uganda are among the countries that have launched programmes to combat this major obstacle to curbing the epidemic. The Bahamas is one of a number of countries that have outlawed discrimination against HIV-infected people in the workplace.

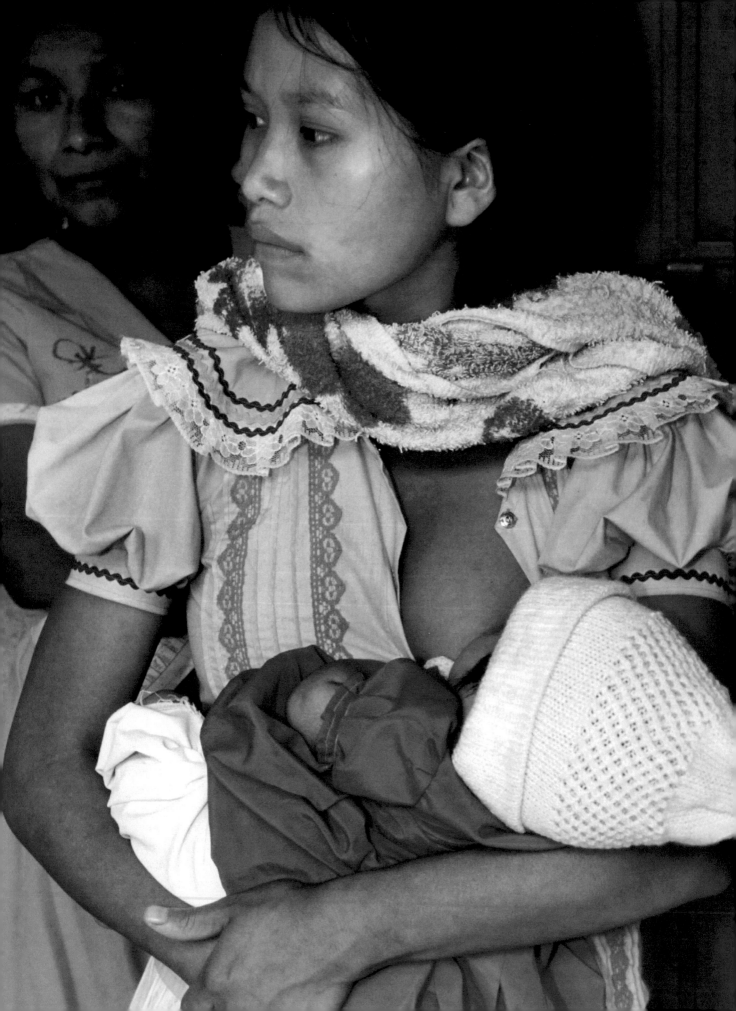

9 Adolescents and Young People

Ensuring the health and well-being of the world's adolescents and young people, equipping them with life skills, and creating educational and employment opportunities for them is a fundamental necessity in meeting the development challenges of the 21st century. The ICPD gave unprecedented attention to adolescents' diverse needs with regard to reproductive health, as both a human rights priority and a practical necessity.

Since 1994 and especially in the past few years, countries have made significant progress in addressing the often-sensitive issues of adolescent reproductive health, including needs for information, education and services that will enable young people to prevent unwanted pregnancy and infection. Increasingly, these efforts are being undertaken as part of a wider, holistic approach that aims to reach young people in diverse situations and equip them to shape their own future.

But enormous challenges remain. One person in five—1.3 billion in all—is an adolescent (defined as ages 10-19), part of the largest youth generation in history. Half are poor, and one fourth live on less than $1 per day. Many are sexually active, often without the power, knowledge or means to protect themselves, or the opportunity to direct their energies to more productive areas of their lives.

Young people (15-24) account for half of all new HIV infections, 2.5 million each year, with girls and young women especially at risk. Despite a trend towards later marriage in much of the world, millions of girls are still expected to marry and begin child-bearing in their teens, often before their bodies are ready.

ADOLESCENTS AND THE MDGs. Enabling youth to delay pregnancy is not only a health and human rights imperative; it is also a key to slowing the continuing momentum of population growth and allowing developing countries to reap the economic benefits that lower fertility can bring, and should be given priority in the global effort to eradicate poverty and achieve the Millennium Development Goals (see Chapter 2).

Investing in young people's health, education and skills development, and allowing girls to stay in school and marry later, are also essential to meeting the MDGs related to gender equality, child mortality, maternal health and HIV/AIDS.

Implementing the ICPD Consensus

The ICPD addressed adolescent reproductive health issues including unwanted pregnancy, unsafe abortion and STIs, including HIV/AIDS, through the promotion of responsible and healthy reproductive and sexual behaviour, including voluntary abstinence, and the provision of appropriate services and counselling specifically suited for that age group. It also aimed to substantially reduce all adolescent pregnancies.[1]

The Programme of Action recognized that poor educational and economic opportunities, gender-based violence, early pregnancy and sexual exploitation increase the vulnerability of adolescents, especially girls, to reproductive health risks. It urged governments and the international community to ensure that all adolescents have access to age-appropriate reproductive health information, education, and services, respecting their right to privacy and confidentiality and ensuring that health provider attitudes or other barriers (laws, regulations, or social customs) do not restrict that access. It also called for parents and families, communities, religious institutions, schools, the mass media and peer groups to be involved in meeting adolescents' reproductive health needs.[2]

ADOLESCENTS' REPRODUCTIVE HEALTH NEEDS

The reproductive health needs of adolescents as a group have been largely ignored to date by existing reproductive health services. The response of societies to the reproductive health needs of adolescents should be based on information that helps them attain a level of maturity required to make responsible decisions. In particular, information and services should be made available to adolescents to help them understand their sexuality and protect them from unwanted pregnancies, sexually transmitted diseases and subsequent risk of infertility. This should be combined with the education of young men to respect women's self-determination and to share responsibility with women in matters of sexuality and reproduction.

—from ICPD Programme of Action, para. 7.41.

NEW BENCHMARKS. The 1999 review of ICPD implementation found that young people's reproductive rights and health needs still remained largely neglected, and were an increasing concern in the face of rising HIV infection rates.[3] New targets were set, including a 25 per cent reduction in HIV prevalence among those aged 15-24, by 2005 in the most-affected countries and by 2010 globally. It was also agreed that the vast majority of those aged 15 to 24 (90 per cent by 2005 and 95 per cent by 2010) should have access to the information, education and services necessary to develop the life skills required to reduce their vulnerability to HIV infection.[4]

Second Generation of Programmes

Drawing on experience since 1994, a comprehensive approach to youth programming has emerged as a global consensus in the past few years. It links reproductive health interventions—including programmes that empower adolescents to delay sexual activity and refuse unwanted relations and to protect themselves if sexually active—to efforts to provide adolescents with choices and options through investments in education, job training and citizenship development. Another priority is to increase the voice and participation of young people in health and development decisions and in the broader life of their communities.

These second-generation adolescent and youth programmes are also giving priority to reaching underserved groups of young people including those who are married, those living in rural areas and poor urban settlements, and those who are not in school (a majority of the adolescent population in many countries).

Efforts in this area show great promise, but need to be massively scaled up to adequately confront the enormous challenges facing the world's youth.

---

29 | **INTEGRATING JOB TRAINING WITH REPRODUCTIVE HEALTH EDUCATION**

The new Multi-Media Centre complex in Cotonou, Benin, bustles with activity. In every room, young people from around the country—nearly 300 in all—are learning how to be print journalists, photographers, radio and TV broadcasters, magazine writers, layout artists, computer graphics experts, web designers, videographers, digital videotape editors, and radio and TV technicians.

Launched by UNFPA in cooperation with the Government, the centre integrates job training with education about preventing HIV/AIDS and unwanted pregnancies, so trainees can also become local advocates for healthier behaviours. The television and 24-hour radio station offer programmes produced by and for youth and have large audiences: 1 million TV viewers and 300,000 daily radio listeners.

Many of the adolescents who frequent the centre are dropouts (in Benin, only 7 per cent of girls and 17 per cent of boys go on to secondary school). Without the centre they would have few options to learn livelihood skills or gain sound information about reproductive health.

The centre is one component of a comprehensive project, Health and Social Services for Adolescents (EAGER), supported by the United Nations Foundation as part of a multi-country initiative on adolescent girls. EAGER also supports youth and leisure centres, youth-friendly health clinics and education, with an emphasis on reducing illiteracy among young women and girls.

UNFPA Global Survey Findings

LAWS AND POLICIES. Over 90 per cent of countries responding to the 2003 UNFPA global survey reported having taken measures to address adolescent reproductive health and rights through policies, laws or programmes.[5] For instance, a law in Panama guarantees the rights of pregnant adolescents to remain in school and receive comprehensive reproductive health care. In Ecuador, a new code on children and adolescents enumerates rights to education, information and reproductive health and integrity.[6] Sierra Leone has established a national youth policy designed to mainstream youth initiatives concerns as central inputs to development policies and programmes. Nepal's current poverty reduction plan puts priority on adolescent health and education.[7]

HEALTH EDUCATION. Nearly all countries have introduced health education, including life skills, into school curricula (primarily in secondary education) and programmes designed for out-of-school youth. Some also report using peer education to reach youth both in and out of school. Many have introduced programmes to reach those not in school through clubs, camps and workshops, and a number are using the mass media to reach a range of youth with reproductive health information.

In Bolivia, with UNFPA support, the Ministry of Health established a programme for indigenous young women that combines access to reproductive health services, literacy skills in Spanish and indigenous languages, and actions to improve self-esteem. UNESCO awarded it the International Literacy Prize in 2000.[8]

SERVICES. Ninety per cent of countries responding have taken action to provide adolescents with access to reproductive health care. Many have established youth-friendly services designed specifically for young people. Most of these are on a small scale and many are run by NGOs. Vast needs remain. Even where services are available, adolescents may face barriers, including lack of information, stigma, family opposition, negative provider attitudes, fear their confidentiality will be violated, and misconceptions about the safety and side effects of contraceptive methods.

TRAINING, LIFE SKILLS AND PARTICIPATION. A number of countries are providing young people with training, employment and life skills education, and most reported actions to promote youth participation in policy and programme development through advisory councils or informal consultation through workshops and dialogue with youth organizations. Costa Rica has launched a groundbreaking initiative to involve young people in developing a national youth policy through a newly established National Youth Council supported by UNFPA.[9]

CONSTRAINTS. Ten years after the ICPD, providing reproductive health information and services for adolescents is still controversial in some countries. There is wide recognition that adolescents need to be empowered to abstain from sex as a personal choice, or to protect themselves from unwanted pregnancies and HIV/AIDS and other STIs if sexually active. Mismatches between needs and care are compounded because adolescents often do not trust health professionals, particularly in crowded government-run clinics, and perceive providers as judgemental and lacking respect for the fundamentals of good care provision, including confidentiality and privacy.

Role of NGOs

NGOs are playing important roles in providing care and information, and in advancing adolescents' reproductive rights. In Indonesia, for example, the Government acknowledges that NGOs are often more advanced in providing services to adolescents. It reports that adolescents and "other people who really understand youth" are managing youth centres developed by the Indonesian Planned Parenthood Association.

In Ethiopia, NGOs are collaborating in providing reproductive health services for adolescents through youth centres and community-based programmes. They are also encouraging youth to undertake self-initiated income-generating activities through small-scale loans.

In Jamaica, the Futures Group International, an international NGO, is collaborating with the Ministry of Health in carrying out a mass media campaign promoting abstinence among all adolescents and educating older ones about safer sex.

In India, the seven NGOs in the Kidavri Network for Adolescent Skills (a mix of religious, social action, social research and humanitarian organizations) serve diverse groups of poor and marginalized adolescents, including street children, and promotes youth participation in decision-making.

Key Health and Development Concerns

Eighty-seven per cent of young people 15-24 live in the developing world.[10] People under age 15 constitute 31 per cent of the population in developing countries and 43 per cent in the least developed.[11] In 2000, adolescents and young people between 10 and 24 were 29 per cent of the population in developing countries and 32 per cent in the least developed, compared to 20 per cent in developed countries.[12]

POVERTY AND GENDER: CYCLES AND IMPACTS.

Young people make up one fourth of the 1 billion people who live below the extreme poverty line of $1 a day.[13] Some 106 million youth live in extreme poverty in South Asia, 60 million in sub-Saharan Africa, 51 million in East Asia and the Pacific, and 15 million in Latin America and the Caribbean. Extreme poverty often prevents adolescents from attending school, perpetuating the cycle of poverty and making this group even more difficult for health, education or youth development programmes to reach.

Youth populations continue to grow in poor countries. The poorer the country, the greater the share young people have in that country's population. Contraceptive use and access to health services increase with education and economic status, as do the age at marriage and first birth.

Illiteracy among those aged 15-24 declined in all regions between 1990 and 2000, but is still substantially higher among females than males, and there has been little progress in reducing that gap since 1990.[14] Girls continue to be faced with limited access to education opportunities, often constrained by traditional gender roles that give priority to educating boys.

EARLY SEXUAL ACTIVITY.

In most of the world, young people are reaching puberty at earlier ages and marrying later than in the past, and premarital sexual relations appear to be increasing.

Data for the late 1990s show that, among young women who were sexually active by age 20, 51 per cent in Africa and 45 per cent in Latin America and the Caribbean initiated sexual activity prior to marriage. By contrast, the corresponding proportion for males was 90 per cent in Africa and 95 per cent in Latin America and the Caribbean.[15] In many developed countries, the onset of sexual activity takes place predominantly prior to marriage for both men and women.

ADOLESCENT PREGNANCY.

The ICPD Programme of Action called on countries to "protect and promote the rights of adolescents to reproductive health education, information and care and greatly reduce the number of adolescent pregnancies".[16] While adolescent pregnancy is declining in many countries, it is still a large concern—especially due to the health risks early pregnancy poses for both the mother and child, and its impact on girls' education and life prospects. Women and girls under 20 still account for 17 per cent of all births in the least-developed countries[17] and for 14 million births worldwide each year.

One woman in three in developing countries gives birth before age 20, ranging regionally from 8 per cent in East Asia to 55 per cent in West Africa.[18]

Pregnancy is a leading cause of death for those aged 15 to 19, with complications of childbirth and unsafe abortion being the major factors. Women aged 15-19 account for at least one fourth of the estimated 20 million unsafe abortions and nearly 70,000 abortion-related deaths each year.

For both physiological and social reasons, mothers aged 15 to 19 are twice as likely to die in childbirth as those in their 20s, and girls under age 15 are five times as likely to die as women in their 20s. Obstructed labour is especially common among young, physically immature women giving birth for the first time. Those who don't die from unrelieved obstructed labour may lose their babies and suffer from fistula, a hole in the birth canal that leaves them incontinent and often social outcasts.

RISK OF STIS AND HIV/AIDS.

Every 14 seconds, a young person is infected with HIV/AIDS. In many settings, the number of new infections among young women is several times that for young men.[19] Two thirds of newly infected young people in sub-Saharan Africa are female. Only a small percentage of young people living with HIV/AIDS know their status. In both developed and less-developed countries, most people who become sexually active at a young age do not know how to protect themselves.

Young women are often unable to negotiate condom use with male partners and may fear violence if they try to do so.

One third of new cases of curable STIs each year—more than 100 million—are among women and men younger than 25.[20] Having another untreated STI significantly increases the risk of HIV infection.

HEIGHTENED RISKS FOR MARRIED ADOLESCENTS.

Despite a global trend towards later marriage (the average age of first marriage among women rose from 21.4 in 1970 to 25.5 in 2000[21]), 82 million girls in developing countries who are now between ages 10 and 17 will be married before their 18th birthday.[22]

Married adolescents often face greater reproductive health risks than those not married. They often face familial and social expectations to begin childbearing right after marriage. Their access to contraceptives is often limited. And many face the risk of STIs or HIV infection from older husbands who may have multiple sexual partners, but negotiating condom use is not an option.

A study in the late 1990s found that contraceptive prevalence among sexually active, unmarried adolescents was more than 30 per cent in seven sub-Saharan African countries (Benin, Cameroon, Cape Verde, Kenya, Nigeria, South Africa and Zambia) and more than 60 per cent in six countries in Latin America and the Caribbean (Bolivia, Brazil, Colombia, Costa Rica, the Dominican Republic and Peru), in both cases much higher than among their married counterparts.[23] Condom use in particular was considerably higher among unmarried adolescents in these countries than among those who were married.[24]

Meeting Young People's Needs

Helping adolescents and young people avoid unintended pregnancy and STIs including HIV requires educational efforts promoting responsible attitudes and healthier sexual behaviour, wider access to youth-

---

30   **LETTING GIRLS ACHIEVE THEIR DREAMS**

"I believe when you learn, you'll reach your dreams, but when you marry too young, you lose the chance to achieve your dreams," says Safa, a 15-year-old girl from Yemen.

Safa is involved in a UNFPA-sponsored Girl Guides programme, one of many partnerships between the Fund and government, social and religious groups that work to raise awareness about the consequences of early marriage for girls, including the health risks of early pregnancy, lost developmental opportunities and limited life options.

Safa and others take part in a range of activities—crafts, sports, arts, and education about reproductive health and social issues. After five years in the programme, she is confident and eager to speak to her friends and family about what she's learned, especially with regard to early marriage. While many of her peers have no choice but to marry young, she has other plans: "I want to be a lawyer—the most famous lawyer in all of Yemen."

---

friendly reproductive health services, training and life skills education and action to promote the rights of women and girls.

UNFPA focuses on ensuring that adolescent reproductive health and rights are included in national agendas and translated into policies and actions with tangible outcomes. The Fund works to promote human rights and gender equality, and to support young people's successful transition to adulthood. Priority is given to reaching the most vulnerable—including those living in poverty or harsh circumstances, married youth, youth living with HIV/AIDS or orphaned because of the epidemic, and street children. UNFPA is also working to address adolescents' health and rights more broadly, by emphasizing holistic programming that addresses a range of needs and is responsive to the broader social, economic and cultural context of young peoples' lives.

Promoting Healthier Behaviour

Providing age-appropriate information about sexuality and preventing pregnancy and infection has been shown to encourage responsible behaviour (including abstinence, delay of sexual initiation and limitation of partners). Besides school-based programmes, efforts are being made to reach out-of-school youth who are often the most vulnerable and at risk.

Since the ICPD, many innovative methods and media have been used, including plays and concerts, mass media messages, sports events, telephone hotlines, and peer counselling and peer education by trained young people. Programmes increasingly focus on giving adolescents life skills as well as imparting information related to sexuality.

Peer education programmes can help young people understand how expectations about gender roles undermine their reproductive health, and can give them confidence to resist those norms. Some, for example, aim to empower young people to refuse sexual relations and assert their right to say "no", as well as to insist on safer sex and the use of condoms if sexually active. Others encourage young men to challenge prevailing notions of male dominance in relationships and tolerance of coercion and sexual violence.

Community participation is also important. In Cambodia, community leaders, teachers, parents and even monks (many of whom are young people themselves) are receiving education on reproductive health issues so their understanding and support for adolescents increases.[25]

Youth-friendly Services

A variety of programme models are being used to provide culturally appropriate, youth-friendly reproductive health services that honour privacy and confidentiality, offer convenient hours and locations, and keep fees affordable, among other features. Promising approaches include peer outreach, mobile clinics, programmes in schools and workplaces, and social marketing of condoms at non-traditional outlets easily accessible to young people.

In Senegal, the Adolescent Girls Project—supported by the United Nations Foundation and implemented by UNFPA and UNICEF—provides 10,000 girls and young women aged 15-24 from poor families with comprehensive reproductive health information and youth-friendly services along with livelihood skills and income-generation activities. The project also serves disabled youth.

Legal Progress

In the past several years, a number of countries have passed laws, drafted new constitutions or approved

| 31 | **AFRICAN YOUTH ALLIANCE** |

The African Youth Alliance (AYA) has become a leader in making comprehensive reproductive health programmes of information and care a reality for young people, primarily those aged 10 to 19. AYA's goal is to reduce the incidence and spread of HIV/AIDS and other STIs and improve overall adolescent reproductive health in Botswana, Ghana, Uganda, and the United Republic of Tanzania. Funded by the Bill & Melinda Gates Foundation, AYA is a collaboration among UNFPA, the Program for Appropriate Technology in Health (PATH), Pathfinder International and local NGOs intended to scale up successful programme approaches.

In Botswana, for example, AYA is helping the Botswana Family Welfare Association and the Family Health Division of the Ministry of Health improve their youth-friendly services. Social marketing efforts and outreach activities aim to increase awareness among youth of new and improved services. In Ghana, an AYA committee has been formed to ensure meaningful participation of the Government in adolescent reproductive health activities. AYA has been working extensively with the Ministry of Health to train health providers in youth-friendly health service delivery.

Nearing the fifth and final year of the programme, AYA is focusing both on building the capacity of its local partners and scaling up current activities to maintain the programme beyond its original timeline. A number of AYA in-country partners have received intensive assistance in financial management and strategic planning.

amendments to legal codes that protect and promote adolescents' rights, including their right to reproductive health care, and aim to eliminate disparities in how boys and girls are treated and valued, within families and by society.

Peru approved legislation guaranteeing a right to education with equal opportunities for girls and boys. Another law seeks to enable rural girls to complete secondary school, and calls for eliminating discrimination against young girls and female adolescents. It also mandates separate health services for women in education facilities. Tunisia has also adopted legislation guaranteeing the right to education—without discrimination based on sex or other factors.

Legislation in Argentina and Panama guarantees pregnant adolescents the right to remain in school. Panama's law establishes pregnant adolescents' right to receive integral health care during pregnancy, childbirth and in the post-partum period (services will be provided for free if young people cannot afford care).

The Ministry of Health will train teachers to advise students on reproductive health and deter discrimination.

Nicaragua passed a comprehensive law on youth development that enumerates the rights of youth (defined as those aged 18 to 30) to reproductive health information, sexual education, and reproductive rights, including access to family planning services and information on STIs, unwanted pregnancy, unsafe abortion and HIV/AIDS.

Key Challenges

SCALING UP. Worldwide, a large number of good programmes have been started since the ICPD to address adolescent reproductive health concerns, but most operate on a relatively small scale. A major challenge is to secure the resources and commitment needed to scale up these programmes.

One organization that has had success in this regard is Action Health Incorporated in Nigeria, whose experiences have helped shape a national reproductive health education programme.[26] Government initiatives are also under way.

Following the ICPD, Mozambique made a commitment to investing in youth. It adopted a multisectoral National Youth Policy that involves different government ministries, NGOs and community organizations in an effort to increase youth participation in policy development and to improve their reproductive health. Designed and developed by youth, the national project, *Geração Biz*, promotes behaviour change and serves a spectrum of adolescent populations, including students and out-of-school youth.

PARTICIPATION AND PARTNERSHIP. Youth participation needs to be institutionalized in programme and policy development processes, and youth must be empowered by these processes. An initiative developed by UNICEF, WHO and UNFPA, Meeting the Development and Participation Rights of Adolescent Girls, strives to put adolescence at the forefront of the development agenda through youth participation in the policy process.

Nicaragua, with the help of UNFPA and UNICEF, has developed and implemented a national youth policy that integrates reproductive health in a broader framework of citizenship, peer education and political

32	**REACHING ADOLESCENT GIRLS IN RURAL BANGLADESH**

In Bangladesh, more than half of all girls marry and begin child-bearing by age 20. UNFPA and UNICEF have teamed together to assist both unmarried adolescents in delaying marriage and married adolescents in knowing their rights. UNICEF's intervention, *Kishori Abjijan*, encourages adolescent leadership and role models and works in partnership with the Government and NGOs (the Population Council, BRAC and the Centre for Mass Education in Science). Girls are active partners and participate in non-traditional livelihood skills programmes such as journalism and photography to enhance their confidence and visibility in the community. UNFPA is supporting efforts to heighten adolescents' awareness about reproductive health rights. Both projects focus on empowering adolescents but they also are helping the Government, families, and communities support the girls' development.

participation. Following a nationwide consultation with adolescents, the Government explicitly integrated the reproductive health needs of adolescents into its Poverty Reduction Strategy Paper, the first country in the world to do so.[27]

With support from Finland, UNFPA is in the process of establishing a youth advisory panel to ensure that its policies integrate young people and address their needs, concerns and aspirations. The panel will include members from both developed and developing countries and will focus on three topics initially: HIV/AIDS, the needs of married adolescents, and the role of culture in adolescent reproductive health.[28]

STRATEGIC APPROACHES. A recent evaluation of UNFPA's and IPPF's contributions to advancing adolescents' health and rights in six programme countries found that more attention to policies, processes and to the strategic use of rights-based and gender-sensitive approaches to programming would have made initiatives undertaken to date more effective.[29] Few efforts are reaching marginalized groups of youth effectively, and more work is needed to make quality reproductive health services available and accessible to young people in general. Findings from the evaluation, which was funded by a number of bilateral donors, will be used in UNFPA's work on behalf of the world's young people in coming years.

10 Reproductive Health for Communities in Crisis

One of the most significant achievements since the ICPD has been greatly increased attention to the reproductive health needs of populations made vulnerable by armed conflict or natural disaster.

A decade ago, humanitarian assistance for populations affected by complex emergencies was generally limited to food, water and sanitation, shelter and protection, and basic health care. More deaths occur worldwide from preventable complications of pregnancy and childbirth than from starvation, but basic materials for safe delivery and emergency obstetric care were rarely included in emergency assistance. The risk of unwanted pregnancy and sexually transmitted infection increases dramatically in displacement camps, but few humanitarian actors in these settings were providing family planning services, post-rape treatment and counselling, or even condoms.

This began to change at the ICPD in 1994, where the Programme of Action specifically addressed the reproductive health needs of displaced persons, and refugee women were invited to speak about their reproductive health needs on an international stage for the first time.

In the mid-1990s, UNFPA, the Office of the United Nations High Commissioner for Refugees (UNHCR), WHO and other partners collaborated in the creation of a comprehensive *Inter-agency Field Manual for Reproductive Health in Refugee Settings*,[1] and agreed on a set of minimum standards for care.

UNFPA has assembled the material resources needed in emergency situations into reproductive health kits, made up of 12 sub-kits including supplies for clean and safe delivery, management of obstetric complications, prevention and management of STIs including HIV/AIDS, and family planning. Since 1996, agencies, organizations and governments have ordered and deployed the kits in more than 50 countries and territories.

RIGHTS APPLY IN EMERGENCIES AS NEEDS ESCALATE.

Women of reproductive age are about 25 per cent of the tens of millions of refugees and persons internally displaced by war, famine, persecution or natural disaster. One in five of these women is likely to be pregnant. Neglecting reproductive health in emergencies has serious consequences, including unwanted pregnancies, preventable maternal and infant deaths, and the spread of STIs including HIV/AIDS.

REPRODUCTIVE HEALTH FOR DISPLACED PERSONS

Migrants and displaced persons in many parts of the world have limited access to reproductive health care and may face specific serious threats to their reproductive health and rights. Services must be particularly sensitive to the needs of individual women and adolescents and responsive to their often powerless situation, with particular attention to those who are victims of sexual violence.

—ICPD Programme of Action, para. 7.11

The ICPD affirmed that the right to reproductive health applies to all people at all times. Effective reproductive health programmes safeguard human rights such as the right to health, to freely decide the number and spacing of children, to information and education, and to freedom from sexual violence and coercion.

Safe Motherhood

Pregnancy and childbirth can be dangerous for women in the best of circumstances. Conflicts or natural disasters put pregnant women at even greater risk because of the sudden loss of medical support, compounded in many cases by trauma, malnutrition or disease or exposure to violence.

When a powerful earthquake struck Bam, Iran, in December 2003, more than 85 per cent of the affected area's health infrastructure and more than half of its health care personnel were lost in less than a minute. The trauma of the catastrophe caused many pregnant women to deliver prematurely or to miscarry.

When recent fighting in Sudan forced more than 100,000 refugees to flee to Chad, pregnant women had to give birth on the roadside and in the middle of the desert. The lack of even the most basic items for safe, clean delivery—soap, a clean razor blade for cutting the umbilical cord, and plastic sheeting to lay on the ground—condemned many women to fatal infections, leaving their children motherless and at risk.[2]

A 2002 study found that complications of pregnancy and childbirth were the leading cause of death among women of childbearing age in war-ravaged Afghanistan. Only 7 per cent of Afghan women who died during childbirth were attended by a skilled health care worker.

As in more stable settings, almost all women who develop pregnancy-related complications can be saved from death and disability if they receive treatment in time. Within 72 hours of the earthquake in Bam, UNFPA helped the Iranian Ministry of Health and Medical Education to procure supplies so pregnant women could deliver safely at home, and to establish temporary emergency obstetric care facilities. In Chad and in other refugee sites, UNFPA works with local partners to establish prenatal support and a referral system for obstetric emergencies. In Afghanistan,

UNFPA responded with emergency supplies and equipment during the acute phase of the crisis, and has contributed to longer-term development as well, rehabilitating a maternal hospital and training health care workers, among other activities.

A recent global evaluation by the Inter-Agency Working Group on Reproductive Health in Refugee Settings found that most refugee sites now offer at least some combination of prenatal care, assisted child delivery, management of obstetric emergencies, and newborn and post-partum care. Maternal mortality ratios in refugee camps in Kenya, Pakistan and the United Republic of Tanzania have been found to be lower than in the host country overall or in the refugees' home countries. While some components of maternal health care—particularly emergency obstetric support—still require a great deal of strengthening, a good start has been made since 1994.

Family Planning

Family planning is often considered to be of secondary concern in an emergency or post-conflict setting. But in a war-torn country like Angola or Sierra Leone, where adequate prenatal care, assisted delivery and emergency obstetric care are not available, as many as one in nine women will die as a result of pregnancy or childbirth over the course of their lives. For women in crisis settings, an unplanned pregnancy can be fatal.

Neglecting family planning can have other serious consequences, including unsafe abortions resulting from unwanted pregnancies, pregnancies spaced too closely together, dangerous pregnancies in women who are too old or too young, and the transmission of STIs including HIV/AIDS.

Maintaining a steady supply of contraceptives can be a major challenge in an emergency. Transportation routes may be cut off, distribution networks dissolved and health facilities destroyed. Existing supplies may fall far short of demand when large numbers of people move into a new location.

Although many women in these settings choose to become pregnant, large numbers who would prefer not to face the difficulties of pregnancy, childbirth or having a baby in a displacement camp have no choice because of lack of access to condoms or other methods of contraception.

Even where services and supplies are available, a number of factors can impede their use. A 2001 assessment by the Women's Commission for Refugee Women and Children found that many Angolan refugees in Zambia were reluctant to use family planning methods, despite their availability. The barriers identified included: resistance by husbands; religious and community beliefs that women should have as many children as they can have; lack of community-based distribution programmes; and the difficulty women have in persuading their partners to use condoms. To promote acceptance of family planning methods, the commission recommended a communications campaign targeting men and the ongoing training of peer educators and community workers.[3]

In emergency settings around the world, UNFPA has supplied free condoms as the first step towards restoring family planning services. When the security situation permits, the Fund conducts rapid assessments to identify family planning needs, and often is able to provide relevant background information on the population, including family planning method preferences. When planning medium- and longer-term programmes, UNFPA and its partners endeavour to involve women,

men and adolescents from the affected populations, to help ensure appropriate, culturally sensitive and effective family planning services.

Sexual and Gender-based Violence

Rape has been a feature of armed conflict for centuries, often employed systematically to humiliate, dominate or disrupt social ties among the "enemy".

In a number of conflicts since ICPD, including those in Bosnia, Rwanda and Kosovo, civilian populations have been deliberately targeted by sexual violence, drawing the attention of the human rights and women's movements and the international press.

Less attention has been given to the women and girls who, during flight and in refugee settings, may be forced to offer sex in exchange for food, shelter or protection. Domestic violence and marital rape also rise significantly among displaced populations, as many men who have lost jobs, status and stability take out their frustrations on their partners.

The impact of violence, especially rape, can be devastating. Physical consequences may include injuries, unwanted pregnancies, sexual dysfunction and HIV/AIDS. Survivors may face exclusion from family life and social isolation. Damage to mental health may include anxiety, post-traumatic stress disorder, depression and suicide. Many survivors will not report rapes and others may feel powerless to do so.

Until recently, there were few attempts to prevent sexual and gender-based violence in times of conflict or displacement. But various initiatives targeting conflict-affected populations have shown that it can be prevented, by:

- Raising awareness about and condemning sexual violence as violations of human rights and a threat to public health;

- Supporting education and information campaigns;

- Promoting safety measures for women in displacement camps, including adequate lighting, security patrols, the safe location of services and facilities, and ensuring that water, fuel, fodder and other provisions can be obtained without having to venture too far;

- Advocating for the enactment and enforcement of laws and policies against sexual and gender-based violence, and providing training for police and judges;

- Involving men to promote behaviour change.

UNFPA supports such efforts, along with treatment and counselling that help create a feeling of safety, and provide opportunities to talk about violent experiences—all of which are vital for recovery. Counselling and education can help family members and communities to accept and support women who have been violated. Training on how to help victims of sexual violence can improve the sensitivity of health workers' responses. Medical and psychological treatment includes emergency contraception, counselling and reproductive health services.

HIV/AIDS and Other STIs

All sexually transmitted infections, including HIV/AIDS, thrive under crisis conditions, which coincide with limited access to the means of prevention, treatment and care.

Other conditions increasing the risk of exposure in emergencies include:

- Large movements of people;

- The break-up of stable relationships and the disintegration of community and family life;

- Disruption of social norms governing sexual behaviour;

- Adolescents starting sexual relations at an earlier age;

- Coercion of women and adolescent girls and boys to exchange sex for food, shelter, income and protection;

- Mixing of populations with higher rates of HIV infection;

- Increased risk of sexual violence, including rape.

Rape by infected men directly exposes women to HIV, and resulting abrasions or tearing of vaginal tissues may increase the risk of infection dramatically.

In some conflicts, the planned and deliberate HIV infection of women has been a tool of ethnic warfare. An association of Rwandan genocide widows found that two thirds of its members who had been raped by Hutu militants were HIV-positive.

While data on HIV prevalence in refugee settings are scarce, it is believed that displaced populations are at increased risk of contracting the virus during and after displacement.[4]

The STI/HIV/AIDS interventions needed in refugee settings, once the situation has stabilized, are much the same as those for settled populations: information and education, condom promotion and distribution, use of syndromic case management for STIs, voluntary counselling and testing for HIV, precautions to ensure safe blood supply, and prevention of mother-to-child transmission.

But in post-conflict settings like Liberia and Sierra Leone, where years of war and continual displacement have created a situation in which growing HIV/AIDS prevalence poses a major threat to post-conflict reconciliation and reconstruction, UNFPA and partners have developed a more comprehensive approach (see Box 28, page 71).

Adolescent Reproductive Health

Young people separated from their families and communities are especially vulnerable to sexual exploitation and are more likely to engage in risky sexual behaviour. War-affected adolescents may be deeply affected by the breakdown of social and cultural systems, loss of access to education and health services, the disruption of school and friendships, exposure to violence and the loss of family members.

In Colombia, violence and displacement have been accompanied by a marked increase in teenage pregnancies and unsafe abortions. One study found that displaced girls were three times as likely as other girls to become pregnant before age 15.[5] In Liberia, where pregnancy among girls as young as 11 and 12 is common, a WHO representative estimated in 2002 that up to 80 per cent of displaced girls had undergone an induced abortion by age 15.[6]

Early pregnancy can have severe implications for the health and well-being of girls whose bodies are not sufficiently developed to withstand pregnancy and childbirth. Girls between ages 10 and 14 are five times more likely to die in pregnancy and childbirth than women between 20 and 24. Unsafe abortions also pose tremendous health risks.[7] In many conflict settings, young girls are extremely vulnerable to HIV and other STIs as well.

One of the most effective ways to protect the health of adolescents affected by disaster is to ensure they have access to sexual and reproductive health information and services. This includes the provision of youth-friendly information and support, and counselling, which can be especially important for victims of sexual violence.

With the support of Belgium, UNFPA is working with local partners to expand services and support for internally displaced youth in Burundi, Colombia, the Democratic Republic of the Congo, Liberia, the

Occupied Palestinian Territory, Rwanda, and Sierra Leone.

In the Democratic Republic of the Congo, for example, UNFPA and a local NGO have established youth centres for displaced young people living outside camps or in big cities. The centres offer reproductive health services including voluntary counselling and testing for STIs. The project has also trained ten NGOs to provide adolescent reproductive health services and information. As demand increases for these services, more funding, resources and partnerships will be needed.

Gains and Gaps

While international funding for reproductive health needs in emergencies has increased since 1994, the number of people requiring these services has grown faster than related assistance. More than half the countries in sub-Saharan Africa have been affected by crisis over the past decade—whether directly, as in Rwanda or Liberia, or indirectly, as in the United Republic of Tanzania and Guinea, which have been burdened by large numbers of refugees from neighbouring countries.

Failure to provide for the reproductive health needs of populations affected by crisis, especially in the age of AIDS, can have tragic consequences not only for individual women, men and children. It can also undermine an entire nation's stability and prospects for post-conflict reconciliation, reconstruction and development.

A new global evaluation by the Inter-Agency Working Group on Reproductive Health in Refugee Settings warns that recent progress in this area is now threatened by stagnant or declining donor funding, compounded by the United States administration's political opposition to some aspects of reproductive health. Increased advocacy and funding are more critical than ever before, as geopolitical instability and increasing vulnerability to natural disasters threaten to increase the number of people in need in coming years.

11 Action Priorities

In adopting the ICPD Programme of Action in 1994, the world's governments recognized that investing in people, broadening their opportunities and enabling them to realize their potential as human beings is the key to sustained economic growth and sustainable development.

Successful action to implement the Cairo agenda and combat poverty depends on adequate funding and effective partnerships.

This chapter discusses the role of partnership in promoting better reproductive health and efforts to achieve the goals of ICPD and the MDGs, and the resources required; this is followed by a summary of priorities for action.

Partnership with Civil Society

Non-governmental organizations, the backbone of many programmes around the world, played a crucial role in shaping the ICPD consensus, and their level of participation in the intergovernmental process was unprecedented. The Programme of Action was far-reaching in its recommendations for promoting partnerships with NGOs, other civil society institutions and the private sector.

In many countries today, NGOs are active in providing reproductive health services and promoting the Cairo agenda in numerous other ways, including advocacy.

Before 1994, partnerships between governments and NGOs mostly involved family planning associations, which had been major providers of family planning services in many developing countries. These collaborations continued over the past decade, with NGOs frequently receiving substantial external funding to provide services independent of governments.

Since the ICPD and its 1999 review, partnerships have developed between governments and a broader range of civil society organizations, including professional associations, community groups and others.

In the UNFPA 2003 global survey, 90 per cent of governments in all regions reported active partnerships on population and reproductive health. Both sides have accepted that NGOs often can reach some groups more easily and carry out certain programmes more effectively than can governments.

NGOS AND REPRODUCTIVE HEALTH SERVICES. Some governments do not provide certain components of reproductive health due to financial constraints or a lack of capacity. In some settings requiring flexibility and quicker outreach, NGOs are better-placed than governments to promote gender equality, address gender-based violence, encourage male responsibility, provide reproductive health information and services to adolescents, undertake youth development programmes, and reach groups at higher risk of HIV infection.

In Mexico, the Government recognizes the role of NGOs in providing medical services including cervical smear tests, gynaecological consultations, antenatal care and care of newborns.

With the introduction over the past decade of health sector reform, Poverty Reduction Strategy Papers (PRSPs) and sector-wide approaches, many donors are now providing funding directly to governments. As a result, developing country governments are now often in a better position to partner with NGOs, other civil society actors and the private sector in ways that complement programmes they are implementing.

In Bangladesh, where NGOs and the private sector provide most health care, the Government has included NGOs and community-based organizations in a National Advisory Committee for Stakeholder Participation in the Health, Nutrition, and Population Sector, to ensure client-focused services, quality care, social and gender equity, and decentralization. The intent is to involve partners in planning as well as implementing policies and programmes.

On the other hand, some countries have included NGOs and others in the development of poverty reduction and health sector reform strategies, but have excluded them from the implementation phase.

NGOs and other civil society actors, including professional associations, are taking up actions traditionally considered the sole province of governments, including setting standards and ensuring accountability, either as a complement to or to fill gaps in government efforts. NGOs can also monitor government compliance with human rights treaties and commitments made to implement the Programme of Action.

Professional associations of doctors, nurses, midwives and other health care staff have key roles to play in standard setting, including ethical standards, and in providing continuing medical education and skills training to their members related to reproductive and sexual health and rights and the elements of quality care. International NGOs like the Commonwealth Medical Trust have carried out activities for this purpose.

PRIVATE SECTOR. Another new development in the past decade has been partnership between NGOs and the private sector in promoting reproductive health, as each side has come to recognize the other's comparative advantages. Private companies provide opportunities for social marketing of condoms and other reproductive health commodities through supermarkets, shops and pharmacies. Some private employers now recognize the benefits of promoting sexual and reproductive health among their employees and in the communities where they operate.

PARLIAMENTARIANS. Elected representatives play important roles in setting priorities, allocating resources and defining institutional responsibilities with regard to sexual and reproductive health services and reproductive rights. Parliamentarians' groups in a number of countries have worked to promote implementation of and adequate funding for the ICPD agenda.

The first such group, the Japan Parliamentarians Federation for Population, has been a leader in the global parliamentarians movement for three decades.

The United Kingdom's All Party Parliamentary Group on Population, Development and Reproductive Health has inspired national parliamentary groups in other European countries. Groups in developing countries include India's Association of Parliamentarians on Population and Development, and committees on population and development in both houses of the Nigerian Parliament.

Regional and global networks of parliamentarians are also active in advocacy efforts. The Asian Forum of Parliamentarians on Population and Development, the Forum of African and Arab Parliamentarians on Population and Development, and the Inter-American Parliamentary Group on Population and Development all scheduled events in 2004 to commemorate the 10th anniversary of the ICPD.

From 18-19 October 2004, the second International Parliamentarians' Conference on the Implementation of the ICPD Programme of Action (IPCI/ICPD) will be held in Strasbourg, France. The conference is jointly organized by the Inter-European Parliamentarian Forum on Population and Development and UNFPA in collaboration with the Council of Europe.

At the first IPCI/ICPD in 2002 in Ottawa, Canada, 103 elected officials from 72 countries signed a Statement of Commitment' outlining specific actions they will take to safeguard women's reproductive rights, improve access to reproductive health services including family planning, reduce maternal mortality and prevent the spread of HIV/AIDS, and pledged to strive to allocate up to 10 per cent of their nations' development budgets for population and reproductive health programmes.

UNIVERSITIES. Governments frequently turn to universities to collect and analyse data, and for research, on sexual and reproductive health issues. In India, for example, 18 Population Research Centres attached to universities are responsible for researching population trends and dynamics, knowledge and attitudes of clients, operational issues and other aspects of the population and development nexus. The Institute of Social, Statistics and Economic Research in the University of Ghana provides training in reproductive health, gender, poverty and population-development interrelationships.

SOUTH/SOUTH COOPERATION. Facilitating the exchange of know-how and experiences between developing countries is another important aspect of the ICPD's emphasis on partnership. Partners in Population and Development, established in 1994 with the support of UNFPA, the Rockefeller Foundation and other donors, is now an alliance of 20 developing countries (Bangladesh, Benin, China, Colombia, Egypt, the Gambia, India, Indonesia, Jordan, Kenya, Mali, Mexico, Morocco, Nigeria, Pakistan, Thailand, Tunisia, Uganda, Yemen and Zimbabwe) working to expand and improve South-South collaboration on family planning and reproductive health.[2]

Resources for Implementing the Programme of Action

The ICPD Programme of Action was the first international consensus document to include estimates of the cost of implementing specified interventions.[3] It defined a basic programme of priority actions to be undertaken in the primary health system, including:

• Family planning and infrastructure for service delivery;

• Additional reproductive health services (including prenatal care, normal and safe delivery; information, education and communication about reproductive health—including STIs, human sexuality and responsible parenthood, and against harmful practices such as female genital cutting—prevention of infertility; counselling, diagnosis and treatment of sexually transmitted infections; and referrals, education and counselling for complications of pregnancy and delivery);

• Prevention of STIs including HIV/AIDS;

• Data, research and policy development for population and reproductive health.

The annual cost of this intervention package was estimated at $17.1 billion in 2000, increasing to $18.5 billion in 2005, $20.5 billion in 2010 and $21.7 billion in 2015. The consensus reached was that developing countries would mobilize two thirds of the require-

ments out of domestic resources and that donor countries would provide international assistance on the order of one third of the total.

The Programme of Action noted that these estimates would be reviewed over time,[4] and that additional resources would be needed at different levels of health systems and for supportive interventions in areas such as education, mortality reduction, women's empowerment and social participation.

NEW PROJECTIONS. After the ICPD, other cost estimates were generated for the broader development agenda. The 20/20 Initiative for meeting basic social service needs, endorsed by the World Summit for Social Development in 1995, called on developing countries to devote 20 per cent of their national budgets for health, education and other social aspects of development, and on donor countries to allocate 20 per cent of their development assistance to these areas. The Commission on Macro-economics and Health estimated total requirements for a priority set of health interventions in low-income countries at $66 billion per year.[5]

Within reproductive health, an estimate of $7-10 billion needed annually for a comprehensive package of HIV/AIDS prevention, treatment and care was presented to the UN General Assembly Special Session on HIV/AIDS in 2001. Subsequent analyses increased that projection.[6] It was estimated that $9.2 billion per year would be needed by 2005 to implement key interventions and develop infrastructure. The prevention components of these estimates were only marginally higher than the ICPD estimate.[7]

New estimates of requirements for the total HIV/AIDS package are now being developed to reflect both the continuing spread of the pandemic and the need for additional funds to strengthen health infrastructure in order to deliver needed services.

This example demonstrates the dynamic nature of resource projections, as intervention priorities are adjusted to local conditions and implementation costs and system requirements are better understood. Costs of the transition to new planning, management and service delivery systems are always hard to anticipate.

Linking HIV/AIDS prevention with reproductive health programmes is a priority policy concern.[8] New vertical programmes addressing HIV/AIDS have start-up costs for dedicated management systems and other institutional requirements.[9] While it is easier to track resource flows for special vertical programmes (rather than having to tease out targeted costs from general health system budgets), integrated programming can address multiple needs and capitalize on synergies between different components, while providing the advantage of economies of scale.[10]

Despite the inherent difficulties in tracking resource flows, UNFPA regularly reports on funding for the ICPD basic population and reproductive health package. Donors in 2003 contributed about $3.1 billion, just 54 per cent of the Programme of Action's donor commitment for 2000 and 51 per cent of the requirement for 2005.

Developing country domestic expenditures for the package in 2003 were estimated at $11.7 billion. However, a large proportion of this outlay comes from a few large countries such as Brazil, China, India, Indonesia and Mexico. Many countries—particularly the poorest, with low per capita expenditures on health—depend mainly on donor funding for their family planning, reproductive health, HIV/AIDS and population-related data, research and policy needs.[11]

The constraints on progress are not only financial. Exchange of information and technology and other forms of technical assistance will be needed so that resources can be most effectively deployed.

HUMAN RESOURCE NEEDS. Progress cannot be accelerated and quality cannot be improved unless programmes can recruit, train and retain staff. Different positions require different skills—technical medical training, counselling abilities, community outreach capacity, supervisory and managerial talents, etc.—and these are frequently in short supply. Civil service salaries are often insufficient to attract the most capable men and women.

Expanding the range of coverage of programmes also requires the ability to ensure that people are available where needs are greatest—often in settings that are remote or lacking amenities. Each of these human resource challenges must be addressed systematically, usually within the context of overall system reform.

COMMODITY NEEDS. Additional progress cannot be made without provision of the essential commodities needed to implement programmes. UNFPA, in collaboration with other major donors, has worked to ensure a reliable supply of quality reproductive health drugs, equipment and supplies.

It is estimated that donors today supply much less than their historical share of contraceptive commodity costs: in the early 1990s, donors provided 41 per cent of commodity requirements, about twice what they provide today. Due to this funding shortfall, systems have had to be developed to handle emergency requests from countries to prevent stockouts and shortages.

MILLENNIUM DEVELOPMENT GOALS. The UN Millennium Project[12] is giving priority to needs-based assessments of resource and capacity requirements to attain the MDGs over the next 11 years. The expert assessments recognize that the availability of reproductive health services (including family planning, safe motherhood and prevention of sexually transmitted infections) is central to achieving the MDGs.[13]

Achieving the MDGs will therefore require multi-sectoral investments, including investments in population and reproductive health.

---

35 | **POPULATION DYNAMICS AND POLICY DEVELOPMENT**

Nicaragua's 2003 National Development Plan is a good example of how countries can integrate population dynamics in the national policy and planning process.

Drafted with technical assistance from UNFPA, the plan notes that population growth and internal and external migration all have important implications for poverty reduction. It calls for improved demographic data collection systems, employment creation, and a special focus on young people's needs for reproductive and sexual health education and services.

The plan has influenced other development processes the Government has participated in or initiated, including the drafting of a Poverty Reduction Strategy Paper, the formulation of the UN Development Assistance Framework, identification of actions needed to meet the MDGs, and national plans on population, youth and development. It has also led to fruitful dialogue among national policy makers in various sectors of development work, and to local collaboration among different sectors, notably in providing adolescent services.

Priorities for Action

Political leadership and adequate funding will be critical to meet both the goals of Cairo and the MDGs. Priorities for action over the next 10 years include:

POLICY COORDINATION

- **Integrate ICPD priorities into development policy dialogues** on poverty eradication, women's empowerment, social policies, human rights, environmental sustainability and macroeconomic polices, and in sector-wide approaches, PRSPs and other programming processes;

- **Broaden policies and programmes to meet the needs of the poorest populations** and ensure that ICPD implementation efforts have a pro-poor orientation. Give priority to increasing the education and skills of the poor, and to providing services to poor rural and urban communities;

- Make **civil society participation** a routine aspect of national, regional and local institutional practices;

- **Reform laws, policies and institutions to promote gender equality and equity.** Combat gender-based violence and harmful traditional practices; expand women's access to land and credit; increase women's participation in decision-making; and redress inequality within families, workplaces and communities;

- **Link national capacity-building efforts** and systems aimed at achieving the MDGs and monitoring progress to those needed to implement the ICPD Programme of Action, to maximize synergy and programme effectiveness.

POPULATION AND DEVELOPMENT

- **Include population dynamics in national planning** and policy dialogues. As population size, composition and density change, planners need to anticipate and meet infrastructure and service needs;

- **Respond to rapid urbanization**, including in the least-developed countries. Expand primary health care—including reproductive health—and other social services in the poor communities on the margins of cities. Facilitate decentralized decision-making by training local staff in budgeting, service delivery and monitoring;

- **Pay more attention to rural development** to: address gaps in health care, education and employment; halt environmental degradation; slow the migration of those with skills and education; and reduce the impact of HIV/AIDS.

REPRODUCTIVE HEALTH

- **Focus more attention and resources on providing comprehensive, high-quality reproductive health services;**

- **Give priority to reproductive health and family planning in efforts to strengthen and reform health systems**, and in sector-wide approaches, PRSPs and strategies for meeting the MDGs;

- **Strengthen capacity** at all levels to provide reproductive health services, ensure sustainable financing and adequate staffing, improve service quality and increase use;

- **Ensure the sustainability and security of supply chains** of all commodities, equipment and supplies needed for comprehensive reproductive health care, including contraceptives;

- **Direct capacity and resources to interventions known to be most effective** including new approaches for reducing maternal mortality and ensuring adolescents' reproductive health;

- **Improve the quality of care**, building on progress in the past decade;

- **Establish effective monitoring and evaluation mechanisms** to address constraints in programme implementation and to assess success;

- **Strengthen systems for data collection, analysis, dissemination and use**, through institutional support and training.

MATERNAL HEALTH

- **Scale up and expand women's access to interventions most effective in preventing maternal deaths**: deliveries with skilled attendants and access to emergency obstetric care;

- **Strengthen family planning services** to enable women to postpone, space and limit pregnancies;

- **Increase the use of safe motherhood services**, particularly among poor households, by raising awareness, addressing social and economic barriers and improving infrastructure;

- **Remove financial obstacles to antenatal, delivery and post-partum care**, by making all such care free or covering fees through national insurance systems.

HIV/AIDS

- **Link HIV/AIDS interventions more effectively to reproductive health care** as part of a multisectoral response to the epidemic;

- **Expand support for family planning and reproductive health programmes**—important entry points for HIV prevention, treatment and care;

- Strengthen efforts to **integrate HIV/AIDS prevention and treatment into comprehensive reproductive health services**;

- **Scale up current programmes**, using a multisectoral approach;

- **Address socio-economic factors facilitating HIV/AIDS infection**, including gender relations, through culturally appropriate behaviour change programmes.

ADOLESCENT REPRODUCTIVE HEALTH

- Intensify efforts to **reach all adolescents in need**, including married adolescents and those not in school;

- **Increase youth participation** in programme design, implementation and monitoring, and in policy processes;

- **Expand the comprehensive approach to youth programming and development**;

- **Scale up current efforts**.

RESOURCES

- **Increase donor assistance** directed to ICPD implementation;

- **Exchange information** on what works, so that funds and administrative capacity combine for maximum impact;

- **Support governments to make good decisions** in an atmosphere of transparency and accountability.

Resources (both from donors and within national allocations) for implementing the ICPD and achieving the MDGs must be significantly increased, but the resources must be used effectively. In the changing institutional landscape of decentralization and devolution of decision-making authority in the social sectors, this will require development of skills and capacity at national and lower administrative levels. Many countries, particularly the poorest, lack sufficient financial and human resources; serious plans are needed to overcome these constraints.

Donors reaffirmed their aspiration to provide international assistance at a level of 0.7 per cent of their gross national product at the Funding for Development conference in Monterrey, Mexico, in 2002. Only five donor countries have reached that level of support.[14]

To attain the MDGs and the critically important goals of the ICPD, and make serious progress in reducing the many dimensions of human poverty, past commitments to development assistance must move from declarations of good intentions to active partnerships and investments.

Conclusion

The ICPD in 1994 gave practical meaning to human-centred development. The Programme of Action acknowledges that investing in people and broadening their opportunities and capabilities is indispensable to achieving sustained economic growth and alleviating poverty.

The Cairo consensus stimulated a global response:

- It facilitated further advances in international understandings about women's health and empowerment that were articulated at the Fourth World Conference on Women in Beijing in 1995;

- It promoted a focus on individuals' opportunities in the development dialogue and put it at the centre of population policies and strategies;

- It was a catalyst for an increased role for civil society organizations in the development dialogue;

- It legitimized international agreements on resource needs to attain development targets.

Through such results, it helped pave the way for the Millennium Development Goals.

In this decade, progress has been made in national, regional and international policies consistent with the ICPD vision. Reproductive health has been advanced in policies and institutions. The diversity of demographic situations around the world has been recognized, and nations have been working to design policies and programmes responsive to felt needs.

Topics that were previously ignored in policy discussions—like harmful traditional practices, gender-based violence, adolescent reproductive health, post-abortion care, the health needs of refugees and people living in emergency situations, the security of supplies of reproductive health and family planning commodities, and the role of culture as a vehicle for advancing basic human rights—are now routinely addressed and acted on.

Much has changed in the world since 1994. The ideological and institutional environment for development initiatives has been dramatically transformed. Decentralization in decision-making, changing balances of public and private responsibilities, new financing mechanisms and budgetary constraints, sector-wide reform efforts, disease-specific vertical programmes and the increased priority given to poverty reduction have transformed the terms of discussion and action.

Yet the person-centred participatory vision of national action at the heart of the ICPD Programme of Action is today more relevant than ever. As the world seeks to reach the ambitious goals of the Millennium Summit, political commitment and the devotion of adequate financial and human resources to implement the ICPD Programme of Action remain centrally important.

Better maternal and child health, gender equality, educational advancement, poverty reduction, environmental quality and improved development partnerships all depend on mobilizing the political will and funding needed to realize the Cairo consensus. Universal access to reproductive health, education and social participation are vital to personal and national dignity, security and progress in alleviating poverty.

Notes and Indicators

Notes

CHAPTER 1

1   UNFPA. 2004. *Investing in People: National Progress in Implementing the ICPD Programme of Action.* New York: UNFPA.

2   This section draws on the analysis in: To Better Communication Project. 2003. *The Road to Global Reproductive Health: Reproductive Health and Rights on the International Agenda: 1968-2003.* New York: To Better Communication Project, coordinated by the media/advocacy working group of the EuroNGOs and sponsored by UNFPA. Web site: www.unfpa.org/upload/lib_pub_file/199_filename_eurongo.pdf, last accessed 9 June 2004.

3   See: UNFPA. 1997. "Reproductive Health and Human Rights." Ch. 1 in: *The State of World Population 1997: The Right to Choose: Reproductive Rights and Reproductive Health.* New York: UNFPA.

4   UNFPA 2004, pp. 34-35.

5   Unsafe abortion is defined as a procedure for terminating an unwanted pregnancy either by persons lacking the necessary skills or in an environment lacking the minimal medical standards or both (based on: WHO. 1994. *Health, Population and Development* [WHO/FHE/94.1]). WHO position paper prepared for the International Conference on Population and Development, Cairo, 5-13 September 1994. Geneva: WHO.

6   WHO, UNICEF, and UNFPA. 2003. *Maternal Mortality in 2000: Estimates Developed by WHO, UNICEF, and UNFPA.* Geneva: WHO; and WHO. 2003. *Reproductive Health: Draft Strategy to Accelerate Progress towards the Attainment of International Development Goals and Targets* (EB113/15 Add.1). Geneva: WHO.

7   UNAIDS. 2004. *2004 Report on the Global AIDS Epidemic: 4th Global Report* (UNAIDS/04.16E). Geneva: UNAIDS.

8   The increase, however, may not be sustained and may largely represent investments in HIV/AIDS programmes rather than in a comprehensive approach to reproductive health.

CHAPTER 2

1   UN Millennium Project. 2004a. "Task Force 3 Interim Report on Primary Education." New York: UN Millennium Project. Web site: www.unmillenniumproject.org/documents/tf3genderinterim.pdf, last accessed 2 June 2004; and UN Millennium Project. 2004b. "Interim Report of Task Force 4 on Child Health and Maternal Health. New York: UN Millennium Project. Web site: www.unmillenniumproject.org/documents/tf4interim.pdf, last accessed 2 June 2004.

2   Hakkert, R., and G. Martine. 2003. "Population, Poverty and Inequality: A Latin American Perspective." Ch. 6 in: *Population and Poverty: Achieving Equity, Equality and Sustainability,* edited by UNFPA. Population and Development Strategies Series. No. 8. New York: UNFPA.

3   Eastwood, R., and M. Lipton. 2001. "Demographic Transition and Poverty: Effects Via Economic Growth, Distribution and Conversion." Ch. 9 in: *Population Matters: Demographic Change, Economic Growth, and Poverty in the Developing World,* edited by N. Birdsall, A. C. Kelley, and S. W. Sinding. 2001. Oxford: Oxford University Press.

4   National Research Council. 1986. *Population Growth and Economic Development: Policy Questions.* Washington D.C.: National Academy Press.

5   RAND Corporation. 2002. "Banking the 'Demographic Dividend': How Population Dynamics Can Affect Economic Growth." *Population Matters* Policy Brief. No. RB-5065-WFHF/DLPF/RF. Santa Monica, California: RAND Corporation; Bloom, D. E., D. Canning, and J. Sevilla. 2003. *The Demographic Dividend: A New Perspective on the Economic Consequences of Population Change.* RAND Population Matters Monograph Series. Santa Monica, California: RAND Corporation; Paes de Barros, R., *et al.* 2001. "Demographic Changes and Poverty in Brazil." Ch. 11 in: *Population Matters: Demographic Change, Economic Growth, and Poverty in the Developing World,* edited by N. Birdsall, A. C. Kelley, and S. W. Sinding. 2001. Oxford: Oxford University Press; Bloom, D. E., D. Canning, and J. Sevilla. 2002. *Demographic Change and Economic Growth: The Importance of Age Structure.* Santa Monica, California: RAND Corporation; Seltzer, J. 2002. *The Origins and Evolution of Family Planning Programs in Developing Countries.* Santa Monica, California: RAND Corporation; Cassen, R. 1994. *Population and Development: Old Debates, New Conclusions.* New Brunswick, New Jersey, and Oxford: Transaction Publishers; and Lassonde, L. 1996. *Coping with Population Challenges.* London: Earthscan Publications.

6   The World Bank. Global Poverty Monitoring web site: www.worldbank.org/research/povmonitor/index.htm, last accessed 18 May 2004.

7   This and the following paragraphs rely heavily on: UNFPA. 2002. *The State of World Population 2002: People, Poverty and Possibilities: Making Development Work for the Poor.* New York: UNFPA.

8   United Nations. 2003. *The HIV/AIDS Epidemic and Its Social and Economic Implications* (UN/POP/MORT/2003/12). New York: Population Division, Department of Economic and Social Affairs, United Nations.

9   National Research Council. 2001. *Preparing for an Aging World: The Case for Cross-National Research.* Washington, D.C.: National Academy Press.

10   UNFPA. 2004. *Investing in People: National Progress in Implementing the ICPD Programme of Action.* New York: UNFPA.

CHAPTER 3

1   United Nations. 1995. *Population and Development,* vol. 1: *Programme of Action adopted at the International Conference on Population and Development: Cairo: 5-13 September 1994,* paragraph 3.14. New York: Department of Economic and Social Information and Policy Analysis, United Nations.

2   Gardner, G., E. Assadourian, and R. Sarin. 2004. "The State of Consumption Today." Ch. 1 in: *State of the World 2004,* by the Worldwatch Institute. 2004. New York: W. W. Norton.

3   WHO. 2002. *World Health Report 2002: Reducing Risks, Promoting Healthy Life,* p. 51. Geneva: WHO.

4   Gardner, Assadourian, and Sarin 2004.

5   Knickerbocker, B. 22 January 2004. "If the Poor Get Richer, Does the World See Progress?" *The Christian Science Monitor.*

6   Gardner, Assadourian, and Sarin 2004.

7   Ibid., p. 5.

8   MacDonald, M., and D. Nierenberg. 2003. "Linking Population, Women, and Biodiversity," p. 42. Ch. 3 in: *State of the World 2003,* by the Worldwatch Institute. 2003. New York: W. W. Norton.

9   UNFPA. 2001. *The State of World Population 2001: Footprints and Milestones: Population and Environmental Change,* pp. 38-39 and the references cited therein; and Nierenberg, D. 2002. *Correcting Gender Myopia, Gender Equity, Women's Welfare, and the Environment,* pp. 29-35. Worldwatch Paper. No. 161. Washington, D.C.: Worldwatch Institute.

10   Nierenberg 2002.

11   Simpson-Hebert, M. 1995. "Water, Sanitation, and Women's Health: The Health Burden of Carrying Water." *The Environmental Health Newsletter* 25. Geneva: WHO. Cited in: *Women, Men, and Environmental Change: The Gender Dimension of Environmental Policies and Programs,* p. 4, by Justine Sass. 2002. Emerging Policy Issues in Population, Health, and the Environment Series. Washington, D.C.: Population Reference Bureau.

12   Nierenberg 2002, p. 45.

13   Women's Environment and Development Organization. 2003. *Common Ground: Women's Access to Natural Resources and the United Nations Millennium Development Goals,* p. 5. New York: Women's Environment and Development Organization.

14   MacDonald and Nierenberg 2003, p. 52; and "Final Conclusions: Meeting of Women Leaders on the Environment: 7-8 March 2002: Helsinki, Finland." Web site: www.mtnforum.org/resources/library/mwlen02a.htm, last accessed 4 May 2004.

CHAPTER 4

1   United Nations. 1995. *Population and Development,* vol. 1: *Programme of Action adopted at the International Conference on Population and Development: Cairo: 5-13 September 1994,* Chapter 9, Section B. New York: Department of Economic and Social Information and Policy Analysis, United Nations.

2   Ibid., paragraphs 9.1 and 9.12.

3   This and subsequent discussions are indebted to the work of Mark R. Montgomery and colleagues, including: Montgomery, M. R. 2004. "Urbanization, Poverty and Health in the Developing World." Presentation at the United Nations Population Division, New York, 8 April 2004; and Montgomery, M., and P. Hewett. 2004. "Urban Poverty and Health in Developing Countries: Household and Neighborhood Effects." Policy Research Division Working Papers. No. 184. New York: The Population Council.

4   Current estimates are taken from: United Nations. 2004. "Executive Summary." *World Urbanization Prospects: The 2003 Revision.* New York: Population Division, Department of Economic and Social Affairs, United Nations.

5   See: Dyson, T. 2003. "HIV/AIDS and Urbanization." *Population and Development Review* 29(3): 427-442.

6   UNFPA. 2004. *Investing in People: National Progress in Implementing the ICPD Programme of Action.* New York: UNFPA. Such a report on recent action is consistent with the finding of the UN Population Division that almost three quarters of developing countries have policies to reduce the migration flow to metropolitan areas, including those adopted earlier. (See: United Nations 2004.)

7   Source: United Nations. 2002. *International Migration Report 2002* (Sales No. E.03.XIII.4). New York: Population Division, Department of Economic and Social Affairs, United Nations.

8   Estimating the number of migrants is difficult, due to a lack of comparable data. While many countries regularly conduct censuses, dissemination of the information generated has been limited or slow, especially for developing countries. Political considerations sometimes influence the reporting of migration statistics in some countries.

9   A long-term migrant is a person who moves to a country other than that of his or her usual residence for a period of at least a year (12 months), so that the country of destination effectively becomes his or her new country of usual residence. From the perspective of the country of departure, the person will be a long-term emigrant and from that of the country of arrival, the person will be a long-term immigrant. A short-term migrant is a person who moves to a country other than that of his or her usual residence for a period of at least three months but less than a year (12 months), except in cases where the movement to that country is for purposes of recreation, holiday, visits to friends and relatives, business, medical treatment or religious pilgrimage. For purposes of international migration statistics, the country of usual residence of short-term migrants is considered the country of destination during the period they spend in it. (See: United Nations. 1998. *Recommendations on Statistics of International Migration: Revision 1* [ST/ESA/STAT/SER.M/58/Rev.1]. Statistical Papers. Series M. No. 58. Rev. 1. New York: Statistics Division, Department of Economic and Social Affairs, United Nations.). According to UNHCR High Commissioner's 20 August 2003 Report to the General Assembly (See: United Nations. 2003. *Report by the High Commissioner to the General Assembly on Strengthening the Capacity of the Office of the High Commissioner for Refugees to Carry Out Its Mandate* [A/AC.96/980]. New York: United Nations.), while a clear distinction between voluntary and forced migration should continue to be made, the problems of refugees and asylum-seekers (who are forced-migrants), will need to be addressed within the wider context of international migration.

10 Although reference is usually made to "countries", "areas" refers here to both countries (for international migration) and cities/villages within countries (for internal migration).

11 The World Bank. 2003. *Global Development Finance 2003: Striving for Stability in Development Finance.* Washington, D.C.: The World Bank.

12 United Nations 1995, paragraph 10.2.

13 Ibid., paragraphs 10.3-10.8.

14 Ogata, S., and A. Sen. 2003. "People on the Move." Ch. 3 in Human Security Now: Commission on Human Security. Final Report of the Commission on Human Security. New York: Commission on Human Security.

15 The General Assembly, in its resolution 54/212 of 22 December 1999, requested the Secretary-General to submit at its fifty-sixth session, ". . . a report that will, *inter alia,* summarize the lessons learned, as well as best practices on migration management and policies, from the various activities relating to international migration and development that have been carried out at the regional and interregional levels . . . ." (See: United Nations. 2000. *Resolution adopted by the General Assembly: 54/212: International Migration and Development* [A/RES/54/212]. New York: United Nations.)

CHAPTER 5

1   United Nations. 1999. *Key Actions for the Further Implementation of the Programme of Action of the International Conference on Population and Development* (A/S-21/5/Add.1), paragraph 48. New York: United Nations.

2   UNFPA. 2004. *Investing in People: National Progress in Implementing the ICPD Programme of Action.* New York: UNFPA.

3   Centre for Development and Population Activities. 2001. *Adolescent Girls in India Choose a Better Future: An Impact Assessment.* Washington, D.C.: Centre for Development and Population Activities.

4   Barnett, B., et al. 1996. *Case Study of the Women's Center of Jamaica Foundation Programme for Adolescent Mothers.* Research Triangle Park, North Carolina: Women's Studies Project, Family Health International; Gayle, H. 2002. "Jamaican Fathers and their Sons: A Tracer Study of the Baby-fathers and Sons of Female Participants of the Women's Centre Foundation of Jamaica in Westmoreland and St. Catherine, with a Focus on Father-son Relationships and Reproductive Health." Report for the Youth.now Project. Kingston, Jamaica: University of the West Indies and the Futures Group; and McNeil, P. 1999. ´Women's Centre: Jamaica: Preventing Second Adolescent Pregnancies by Supporting Young Mothers." FOCUS on Young Adults Project Highlights. Web site: www.fhi.org/en/Youth/YouthNet/Publications/FOCUS/ProjectHighlights/womenscentrejamaica.htm, last accessed 6 May 2004.

5   Instituto Mexicano de Investigación de Familia y Población (IMIFAP). 2002. *Salud y Empoderamiento para Las Mujeres del Medio Rural y Sus Familias.* IMIFAP Action-Research Summary. Mexico City: Instituto Mexicano de Investigación de Familia y Población. Web site: www.imifap.org.mx/espanol/resumenes/resumen13.pdf, last accessed 6 May 2004; and Pick, S. 2002. "Si Yo Estoy Bien, Mi Familia También (If I Am OK, My Family Is Too)." Paper presented at the PsychoSocial Workshop, Atlanta, Georgia, 7 May 2002.

6   Boland, R. 2004. Population and Law database. Special compendium provided on request.

7   UNFPA 2004, p. 26.

8   Ibid., p. xiii, Chs. 3, 4 and 9.

9   Ibid., p. xxiii.

10 UN Millennium Project. 2004. "Task Force 3 Interim Report on Gender Equality." New York: UN Millennium Project.

11 UNDP, 2003. *Millennium Development Goals: National Reports: A Look through a Gender Lens,* p. 22. New York: UNDP.

12 UN Millennium Project 2004.

13 Commonwealth Secretariat. 1999-2004. GMS Series of Reference Manuals on Gender Mainstreaming in Various Sectors and Development Issues for Policy Makers and Other Stakeholders. London: Commonwealth Secretariat. Web site: www.thecommonwealth.org/gender, last accessed 21 June 2004.

CHAPTER 6

1   Singh, S., et al. 2004. *Adding It Up: The Benefits of Investing in Sexual and Reproductive Health Care.* Washington, D.C., and New York: The Alan Guttmacher Institute and UNFPA.

2   UNFPA. 2004a. *Investing in People: National Progress in Implementing the ICPD Programme of Action.* New York: UNFPA; and WHO. 2003. *Reproductive Health: Draft Strategy to Accelerate Progress towards the Attainment of International Development Goals and Targets* (EB113/15 Add.1). Geneva: WHO.

3   United Nations. 2004. *Review and Appraisal of the Progress made in Achieving the Goals and Objectives of the Programme of Action of the International Conference on Population and Development: Report of the Secretary-General* (E/CN.9/2004/3). New York: United Nations.

4   In Bangladesh, prevalence increased by 1.8 percentage points overall but 3 quintiles—to the largest extent in the poorest quintile—experienced decreases in desires satisfied by modern method use.

5   Details about the calculations, and an extended discussion of the reliability, validity and implications of the concept, can be found in: Casterline, J. B., and S. W. Sinding. 2000. "Unmet Need for Family Planning in Developing Countries and Implications for Population Policy." *Population and Development Review* 26(4): 691-723.

6   United Nations. 1995. *Population and Development,* vol. 1: *Programme of Action adopted at the International Conference on Population and Development: Cairo: 5-13 September 1994,* paragraph 7.16. New York: Department of Economic and Social Information and Policy Analysis, United Nations.

7   United Nations. 1999. *Key Actions for the Further Implementation of the Programme of Action of the International Conference on Population and Development* (A/S-21/5/Add.1), paragraph 58. New York: United Nations.

8   Reports of total and wanted fertility in most recent surveys obtained from: Demographic and Health Surveys' StatCompiler. Web site: www.orcmacro.org, accessed 8 March 2004.

9   The dynamics of this relationship, and the corresponding impact on recourse to abortion are analyzed in: Bongaarts, J. 1997. *Trends in Unwanted Childbearing in the Developing World.* Policy Research Division Working Paper. No. 98. New York: The Population Council; and Bongaarts, J., and C. F. Westoff. 2000. "The Potential Role of Contraception in Reducing Abortion." *Studies in Family Planning* 31(3): 193-202.

10 Singh, S., et al. 2004.

11 Modern contraceptive techniques include male and female sterilization, oral contraceptives, implants and injections and barrier methods (male and female condoms and diaphragms). Traditional methods include periodic abstinence, withdrawal and lactational amenorrhea (extended breast-feeding).

12 Adding those using traditional methods into those with unmet need for modern methods, fully 63 per cent of sub-Saharan women and couples have unmet limiting and spacing desires.

13 This can include shortages of support for and supply of temporary

methods for birth spacing, the need to address cultural sensitivities (e.g., bleeding or spotting side effects where blood taboos are prevalent).

14 Rudy, S., et al. 2003. "Improving Client-Provider Interaction." *Population Reports*. Series Q. No. 1. Baltimore, Maryland: The INFO Project, Center for Communication Programs, the Johns Hopkins Bloomberg School of Public Health. Web site: http://www.infoforhealth.org/pr/q01/q01.pdf, last accessed 27 April 2004.

15 See the discussion of Youth Friendly Services in: UNFPA. 2003. *The State of World Population 2003: Making 1 Billion Count: Investments in Adolescents' Health and Rights*. New York: UNFPA.

16 Calculated from data provided by John Ross from the 1999 Family Planning Program Strength survey (See: Ross, J., and J. Stover. 2000. *Effort Indices for National Family Planning Programs: 1999 Cycle*. Measure *Evaluation* Working Paper. No. WP-00-20. Chapel Hill, North Carolina: Carolina Population Center, University of North Carolina. See also: Ross, J., J. Stover, and A. Willard. 1999. *Profiles for Family Planning and Reproductive Health Programs: 116 Countries*, Ch. 5. Glastonbury, Connecticut: The Futures Group International.). This survey has been conducted periodically since 1982. An update is scheduled for late 2004. Data on unmarried women lacking access to contraceptives are less available, but would add significant numbers to this estimate.

17 Bongaarts, J., and S. C. Watkins (1996). "Social Interactions and Contemporary Fertility Transitions. *Population and Development Review* 22(4): 639-682; and Merrick, T. W. 2002. "Population and Poverty: New Views on an Old Controversy." *International Family Planning Perspectives* 28(1): 41-46.

18 Merrick 2002.

19 The term is used here to denote those who are using a method when they wish to avoid a pregnancy. More complete definitions take into account whether or not users have access to a range of choice and have selected a method appropriate to their desires (see: Sinding and Casterline 2000; Jain, A. K. 2001. "Family Planning Programs: Quality of Care." In: *The International Encyclopedia of the Social and Behavioral Sciences*, edited by N. J. Smelser and P. B. Baltes. 2001. Amsterdam: Elsevier; and RamaRao, S., and R. Mohanam. 2003. "The Quality of Family Planning Programs: Concepts, Measurements, Interventions, and Effects." *Studies in Family Planning* 34[4]: 227-248). An important historical contribution focused on

attainment of intentions can be found in: Jain, A., and J. Bruce. 1994. "A Reproductive Health Approach to the Objectives and Assessment of Family Planning Programs." Pp. 192-208 in: *Population Policies Reconsidered: Health, Empowerment, and Rights*, edited by G. Sen, A. Germain, and L. Chen. 1994. Cambridge, Massachusetts: Harvard University Press.

20 Merrick, T. 2004. "Maternal-Newborn Health and Poverty." Draft. Washington, D.C.: The World Bank. Cited in: UNFPA 2004a.

21 UNFPA 2004a.

22 WHO. 1998. *Emergency Contraception: A Guide for Service Delivery* (WHO/FRH/FPP/98.19). Geneva: Family Planning and Population, Reproductive Health Technical Support, Family and Reproductive Health, WHO.

23 WHO. n.d. "Sexually Transmitted Infections: A Persistent Public Health Burden." Geneva: Department of Reproductive Health and Research, WHO.

24 WHO 2003.

25 WHO n.d.

26 UNFPA. 2004b. "Sexually Transmitted Infections: Breaking the Cycle of Transmission," p. 16. Draft. New York: Reproductive Health Branch, Technical Support Division, UNFPA.

27 United Nations 1995, paragraph 7.32.

28 WHO n.d.

29 Coggins, C., and A. Heimburger. 2002. "Sexual Risk, Sexually Transmitted Infections, and Contraceptive Options: Empowering Women in Mexico with Information and Choice," pp. 274-275. Ch. 15 in: *Responding to Cairo: Case Studies of Changing Practice in Reproductive Health and Family Planning*, edited by N. Haberland and D. Measham. 2002. New York: The Population Council.

30 WHO n.d.

31 WHO 2003.

32 UNFPA 2004b, pp. 25-26.

33 Ibid., pp. 26-27.

34 UNFPA 2004a, p. 37.

35 Ibid.

36 The World Bank. 2004. *World Development Report 2004: Making Services Work for Poor People*, pp. 1-5. New York: Oxford University Press.

37 Bruce, J. 1990. "Fundamental Elements of the Quality of Care: A Simple Framework." *Studies in Family Planning* 21(2): 61-91.

38 Huezo, C., and S. Diaz. 1993. "Quality of Care in Family Planning: Clients' Rights and Providers' Needs." *Advances in Contraception* 9(2): 129-139.

39 John Snow, Inc. 2000. *Mainstreaming Quality Improvement in Family Planning and Reproductive Health Service Delivery: Context and Case Studies*. Arlington, Virginia: Family Planning Expansion and Technical Support (SEATS II) Project, John Snow, Inc.; Lynam, P., L. M. Rabinowitz, and M. Shobowale. 1993. "Using Self-Assessment to Improve the Quality of Family Planning Clinic Services." *Studies in Family Planning* 24(4): 252-260; and Hardee, K., and B. Gould. 1993. "A Process for Service Quality Improvement in Family Planning." *International Family Planning Perspectives* 19(4): 147-152.

40 Hardee, K. Forthcoming. "The Intersection of Access, Quality of Care and Gender in Reproductive Health and STI/HIV Services: Evidence from Kenya, India and Guatemala." Washington, D.C.: Interagency Gender Working Group and the POLICY Project.

41 Mensch, B., M. Arends-Kuenning, and A. Jain. 1996. "The Impact of the Quality of Family Planning Services on Contraceptive Use in Peru." *Studies in Family Planning* 27(2): 59-75.

42 Mroz, T. A., et al. 1999. "Quality, Accessibility, and Contraceptive Use in Rural Tanzania." *Demography* 36(1): 23-40.

43 Pariani, S., D. M. Heer, and M. D. Van Arsdol, Jr. 1991. "Does Choice Make a Difference to Contraceptive Use: Evidence from East Java." *Studies in Family Planning* 22(6): 384-390.

44 Cotton, N., et al. 1992. "Early Discontinuation of Contraceptive Use in Niger and the Gambia." *International Family Planning Perspectives* 18(4): 145-149.

45 Quality of care was defined as the fieldworker usually or always being responsive to a client's questions, appreciating her need for privacy, being perceived as dependable to help with problems, sympathetic to the client's needs, providing enough information, spending 10 minutes or more with the client during the past visit, and providing a choice of methods.

46 Koenig, M. 2003. *The Impact of Quality of Care On Contraceptive Use: Evidence from Longitudinal Data from Rural Bangladesh*. Baltimore, Maryland: Department of Population and Family Health Sciences, Bloomberg School of Public Health, Johns Hopkins University.

47 Rudy, S., et al. 2003.

48 John Snow, Inc. 2000; Nguyen, M. T., et al. 1998. "Improving Quality and Use of Family Planning in Three Sites in Vietnam." Paper presented at the American Public Health Association Annual Meeting, Washington, D.C., 15-19 November 1998; Bradley, J., et al. 1998. *Quality of Care in Family Planning Services: An Assessment of Change in Tanzania 1995/6 to 1996/7*. New York: AVSC International. All cited in: RamaRao, S., and R. Mohanam. 2003. "The Quality of Family Planning Programs: Concepts, Measurements, Interventions, and Effects." *Studies in Family Planning* 34(4): 227-248.

49 RamaRao and Mohanam 2003.

50 Schuler, S. R., L. M. Bates, and M. D. K. Islam. 2002. "Paying for Reproductive Health Services in Bangladesh: Intersections between Cost, Quality and Culture." *Health Policy and Planning* 17(3): 273-280.

51 UNFPA. 2004c. "Donor Support for Contraceptives and Condoms for STI/HIV Prevention 2002." Draft. New York: UNFPA.

52 The average cost per user of contraceptives has been calculated at $1.52 a year. It is assumed that the unavailability of contraceptives would have an adverse impact on women's reproductive health, even where other reproductive health services exist. To estimate the consequences, we have used the formulas employed in: UNFPA. 1997. *Meeting the Goals of the ICPD: Consequences of Resource Shortfalls up to the Year 2000: Report of the Executive Director* (DP/FPA/1997/12). New York: UNFPA.

53 UNFPA. 2002. *Reproductive Health Essentials: Securing the Supply: Global Strategy for Reproductive Health Commodity Security*, Chapter 1. New York: UNFPA.

54 See: UNFPA. 2001. *Reproductive Health Commodity Security: Partnerships for Change: A Global Call to Action*. New York: UNFPA. Web site: www.unfpa.org/upload/lib_pub_file/135_filename_rhcstrategy.pdf, last accessed 17 June 2004.

55 WHO and UNFPA. 2002. "Essential Drugs and Other Commodities for Reproductive Health Services." Draft. Geneva and New York: WHO and UNFPA.

56 United Nations. 2003. *World Population Prospects: The 2002 Revision*. New York: Population Division, Department Economic and Social Affairs, United Nations; and United Nations. 2002. *World Urbanization Prospects: The 2001 Revision*. New York: Population Division, Department Economic and Social Affairs, United Nations.

57 Malhotra, A., and R. Mehra. 1999. *Fulfilling the Cairo Commitment:*

Enhancing Women's Economic and Social Options for Better Reproductive Health. Washington, D.C.: International Center for Research on Women.

58 Greene, M. E., and A. E. Biddlecom. 2000. "Absent and Problematic Men: Demographic Accounts of Male Reproductive Roles." *Population and Development Review* 26(1): 81-115.

59 Ezeh, A. C., M. Seroussi, and H. Raggers. 1996. *Men's Fertility, Contraceptive Use, and Reproductive Preferences*. Demographic and Health Surveys Comparative Studies. No. 18. Calverton, Maryland: Macro International.

60 Research exploring this topic included: Mason, K. O., and A. M. Taj. 1987. "Differences Between Women's and Men's Reproductive Goals in Developing Countries." *Population and Development Review* 13(4): 611-638; and Coombs, L. C., and M. C. Chang. 1981. "Do Husbands and Wives Agree: Fertility Attitudes and Later Behaviour." *Population and Environment* 4(2): 109-127.

61 The Alan Guttmacher Institute. 2003. *In Their Own Right: Addressing the Sexual and Reproductive Health Needs of Men Worldwide*. New York: The Alan Guttmacher Institute.

62 Ibid.

63 Examples of this orientation can be found in context-specific pro-grammes and research in Latin America (Loaiza, E. 1998. "Male Fertility, Contraceptive Use, and Reproductive Preferences in Latin America: The DHS Experience." Paper prepared for the seminar, "Men, Family Formation and Reproduction", organized by the Committee on Gender and Population of the International Union for the Scientific Study of Population [IUSSP] and the Centro de Estudios de Poblacion [CENEP], Buenos Aires, Argentina, 13-15 May 1998. Liege, Belgium: IUSSP) and in Ghana (Lamptey, P. et al. 1978. "An Evaluation of Male Contraceptive Acceptance in Rural Ghana." *Studies in Family Planning* 9(8): 222-226.).

64 Basu, A. M. 1996. "Women's Education, Marriage And Fertility: Do Men Really Not Matter?" Population and Development Program Working Paper Series. No. 96.03. Ithaca, New York: Cornell University; Hull, T. H. 1999. "Men and Family Planning: How Attractive is the Programme of Action?" Paper presented at the Psychosocial Workshop, New York, New York, 23-24 March 1999; and Hawkes, S. 1998. "Providing Sexual Health Services for Men in Bangladesh." *Sexual Health Exchange* 3: 14-15.

65 White, V., M. Greene, and E. Murphy. 2003. "Men and Reproductive Health Programs: Influencing Gender Norms." Washington, D.C.: The Synergy Project. Available at www.synergyaids.com/ SynergyPublications/Gender_Norms. pdf, accessed 5 March 2004.

66 Rivers, K., and P. Aggleton. 2001. *Working with Young Men to Promote Sexual and Reproductive Health*. London: Safe Passages to Adulthood, University of London.

67 Brady, M., and A. B. Khan. 2002. *Letting Girls Play: The Mathare Youth Sports Association's Football Program for Girls*. New York: The Population Council.

68 See: Estudos e Comunicaçao em Sexualidade e Reproducao Humana (ECOS) web site: www.ecos.org.br/, accessed 3 March 2004.

CHAPTER 7

1 Save the Children. 2001. *Behind Every Healthy Child is a Healthy Mother*. Report of the Symposium on the Linkages Between Maternal Health, Family Planning, and Child Survival. Washington, D.C., 24 July 2001. Web site: www.savethechildren.org/ publications/reproductive_health.pdf, last accessed 7 May 2004.

2 WHO, UNICEF, and UNFPA. 2003. *Maternal Mortality in 2000: Estimates Developed by WHO, UNICEF, and UNFPA*. Geneva: WHO.

3 Safe Motherhood Initiative. 2003. Web site: http://www.safemotherhood.org, accessed 22 February 2004. The wide range of 30 to 50 reflects the difficulties of defining and measuring maternal morbidity. Underreporting, misclassification and failure to recognize the condition are common due to social and cultural factors, the nature of the conditions and the resources available to assess them.

4 United Nations. 1995. *Population and Development*, vol. 1: *Programme of Action adopted at the International Conference on Population and Development: Cairo: 5-13 September 1994*, paragraph 8.22. New York: Department of Economic and Social Information and Policy Analysis, United Nations.

5 Freedman, L., et al. 2004. "Interim Report of Task Force 4 on Child Health and Maternal Health," p. 54. New York: UN Millennium Project.

6 Liljestrand, J. 2000. "Strategies to Reduce Maternal Mortality Worldwide." *Current Opinion in Obstetrics and Gynecology* 12(6): 513-517.

7 Starrs, A. 1998. *The Safe Motherhood Action Agenda: Priorities for the Next Decade*, p. 9. New York: Family Care International.

8 Liljestrand 2000.

9 UNFPA. 2004a. *Saving Mothers' Lives: The Challenge Continues*. Brochure. New York: UNFPA.

10 United Nations. 1999. *Key Actions for the Further Implementation of the Programme of Action of the International Conference on Population and Development* (A/S-21/5/Add.1), paragraph 62(a) and 62(b). New York: United Nations.

11 United Nations. 2004. *Review and Appraisal of the Progress Made In Achieving the Goals and Objectives of the Programme of Action of the International Conference on Population and Development: Report of the Secretary-General* (E/CN.9/2004/3). New York: United Nations.

12 WHO. 2003a. *International Statistical Classification of Disease and Related Health Problems*, 10th rev. Geneva: WHO.

13 Maine, D., and T. McGinn. 1999. "Maternal Mortality and Morbidity." Ch. 31 in: *Women and Health*, edited by M. Goldman and M. Hatch. 1999. San Diego, California: Academic Press.

14 WHO, UNICEF, and UNFPA 2003.

15 Graham, W., et al. 2004. "The Familial Technique for Linking Maternal Death with Poverty." *The Lancet* 363(9402): 23-27.

16 Kunst, A. E., and T. Houweling, "A Global Picture of Poor-rich Differences in the Utilisation of Delivery Care." Pp. 297-315 in: *Safe Motherhood Strategies: A Review of the Evidence*, by V. De Brouwere and W.Van Lerberghe. 2001. Studies in Health Services Organisation and Policy Series. No 17. Antwerp: ITG Press.

17 WHO. 2004. Personal communication on forthcoming publications.

18 Maine and McGinn 1999.

19 Fortney, J., and J. Smith, J. 1996. *The Base of the Iceberg: Prevalence and Perceptions of Maternal Morbidity in Four Developing Countries*. Research Triangle Park, North Carolina: Maternal and Neonatal Health Center, Family Health International.

20 Donnay, F., and L. Weil. 2004. "Obstetric Fistula: The International Response." *The Lancet* 363(9402): 71-72.

21 Gay, J., et al. 2003. *What Works: A Policy and Program Guide to the Evidence on Family Planning, Safe Motherhood, and STI/HIV/AIDS Interventions: Module 1: Safe Motherhood*. Washington, D.C.: The POLICY Project.

22 *Saving Mother's Lives: What Works: Field Guide for Implementing Best Practices in Safe Motherhood*. 2002. Washington, D.C.: The White Ribbon Alliance for Safe Motherhood/India.

23 Global Health Council. 2002. *Promises to Keep: The Toll of Unintended Pregnancies on Women's Lives in the Developing World*. Washington, D.C.: Global Health Council.

24 WHO, UNICEF, and UNFPA 2003.

25 Ibid.

26 Fortney, J., and J. Smith. 1999. "Measuring Maternal Morbidity." Pp. 43-50 in *Safe Motherhood Initiatives: Critical Issues*, edited by M. Berer and T. K. S. Ravindran. 1999. Oxford, United Kingdom: Blackwell Science for Reproductive Health Matters.

27 UNFPA. 2004b. *Investing in People: National Progress in Implementing the ICPD Programme of Action 1994-2004*, p. 46. New York: UNFPA.

28 Pathmanathan, I., et al. 2003. *Investing in Maternal Health: Learning from Malaysia and Sri Lanka*. Human Development Network. Health, Nutrition and Population Discussion Paper. Washington D.C.: The World Bank.

29 Ibid.

30 UNFPA 2004b, p. 45.

31 Materials provided by the UNFPA Division of Latin America and the Caribbean.

32 WHO and UNICEF. 2003. *Antenatal Care in Developing Countries: Promises, Achievements and Missed Opportunities : An Analysis of Trends, Levels, and Differentials: 1990-2001*. Geneva and New York: WHO and UNICEF.

33 Liljestrand, J. 1999. "Commentary: Reducing Perinatal and Maternal Mortality in the World: The Major Challenges." *British Journal of Obstetrics and Gynaecology* 106(9): 877-880.

34 UNFPA 2004b, p. 46.

35 Graham, W., J. S. Bell, and H. W. Bullough. 2001. "Can Skilled Attendance at Delivery Reduce Maternal Mortality in Developing Countries?" Pp. 97-129 in: *Safe Motherhood Strategies: A Review of the Evidence*, by V. De Brouwere and W.Van Lerberghe. 2001. Studies in Health Services Organisation and Policy Series. No 17. Antwerp: ITG Press.

36 Evidence regarding the effect of skilled attendance at delivery is complicated by the different definitions used by countries in the training and regulation of midwives.

37 WHO, UNFPA, UNICEF, and the World Bank. 1999. *Reduction of Maternal Mortality: A Joint WHO/UNFPA/UNICEF/World Bank Statement*. Geneva: WHO.

38 Cunningham, F. G., et al. 1993. *Williams Obstetrics*, 19th Edition. Norwalk, Connecticut: Appleton & Lange. Cited in: "Background Paper of the Millennium Project Task Force on Child Health and Maternal Health," by L. Freedman, et al. 2003. New York: United Nations Millennium Project.

39 Graham, Bell, and Bullough 2001.

40 UNFPA 2004b, p. 45.

41 UNICEF, WHO, and UNFPA. 1997. *Guidelines for Monitoring and Availability and Use of Obstetric Services*. New York: UNICEF.

42 Averting Maternal Death and Disability (AMDD). 2003. *AMDD Notebook*, p. 7. No. 8. New York: Averting Maternal Death and Disability, Mailman School of Public Health, Columbia University.

43 UNFPA 2004b, p. 46.

44 UNFPA 2004a.

45 Cholil, A., M. B. Iskandar, and R. Sciortino. 1998. *The Life Saver: The Mother Friendly Movement in Indonesia*. Jakarta, Indonesia: The State Ministry for the Role of Women and the Ford Foundation.

46 Lalonde, A. B., et al. 2003. "Averting Maternal Death and Disability: The FIGO Save the Mothers Initiative: The Uganda-Canada Collaboration." *International Journal of Gynecology and Obstetrics* 80(2): 204-212.

47 WHO. 2003b.

48 UNICEF, WHO, and UNFPA 1997.

49 United Nations 1995, paragraph 8.25.

50 See: Hardee, K., et al. Forthcoming. *What Works: A Policy and Program Guide to the Evidence on Family Planning, Safe Motherhood, and STI/HIV/AIDS Interventions: Module 2: Postabortion Care*. Washington, D.C., and Baltimore, Maryland: POLICY Project, the Futures Group; FRONTIERS Program, the Population Council; and INFO Project, Center for Communication Programs, Johns Hopkins Bloomberg School of Public Health.

51 "Appendix E: Postabortion Care in Kenya: Case Study," p. 12 in: *Global Evaluation of USAID's Postabortion Care Program*," by L Cobb, et al. 2001. Washington, D.C.: Poptech.

52 Postabortion Care Consortium Community Task Force. 2002. "Essential Elements of Postabortion Care: An Expanded and Updated Model." *PAC in Action*, No. 2. Special Supplement.

53 Billings, D., J. Fuentes Velásquez, and R. Pérez-Cuevas. 2003. "Comparing the Quality of Three Models of Postabortion Care in Public Hospitals in Mexico City." *International Family Planning Perspectives* 29(3): 112-120; Johnson, B., et al. 2002. "Reducing Unplanned Pregnancy and Abortion in Zimbabwe through Postabortion Contraception." *Studies in Family Planning* 33(2): 195-202; Medina, R., et al. 2001. *Expansion of Postpartum/Postabortion Contraception in Honduras*. FRONTIERS Program Final Report. Washington, D.C.: The Population Council; and Lema, V., and V. Mpanga. 2000. "Post-abortion Contraceptive Acceptability in Blantyre, Malawi." *East African Medical Journal* 77(9): 488-493.

54 Huntington, D., and L. Nawar. 2003. "Moving from Research to Program: The Egyptian Postabortion Care Initiative." *International Family Planning Perspectives* 29(3): 121-125.

55 Megied, A., and A. Hassan. 2003. "Decentralization of Post-abortion Care to District Hospitals and Rural Health Units." Paper presented at the 17th International Federation of Obstetrics and Gynecology (FIGO) World Congress of Gynecology and Obstetrics, Santiago, Chile, 2-7 November 2003.

56 Htay, T. T., J. Sauvarin, and S. Khan. 2003. "Integration of Post-Abortion Care: The Role of Township Medical Officers and Midwives in Myanmar." *Reproductive Health Matters* 11(21): 27-36.

57 Gebreselassie, H., and T. Fetters. 2002. *Responding to Unsafe Abortion in Ethiopia: A Facility-based Assessment of Postabortion Care Services in Public Health Sector Facilities in Ethiopia*. Chapel Hill, North Carolina: Ipas.

58 Solo, J., et al. 1998. "Creating Linkages Between Incomplete Abortion Treatment and Family Planning Services in Kenya: What Works Best." Operations Research Technical Assistance Africa Project II. Paper Presented at the Global Meeting, Advances and Challenges in Postabortion Care Operations Research. New York: The Population Council.

59 The Population Council. 2000. *Meeting Women's Health Care Needs After Abortion*. Frontiers in Reproductive Health Program Brief. No. 1. Washington, D.C.: Frontiers in Reproductive Health, the Population Council.

60 Settergren, S., et al. 1999. *Community Perspectives on Unsafe Abortion and Postabortion Care: Bulawayo and Hwange Districts, Zimbabwe*. Washington, D.C.: POLICY Project, the Futures Group International.

61 See: Gay, J., et al., 2003.

62 Castro, R., et al. 2000. "A Study of Maternal Mortality in Mexico Through a Qualitative Approach." *Journal of Women's Health and Gender-Based Medicine* 9(6): 679-690.

63 Seone, G., V. Kaune, and V. Cordova. 1996. *Diagnostico: Barreras y Viabilizadores en la Atencion de Complicaciones Obstetricas y Neonatales*. La Paz, Bolivia: MotherCare Bolivia, John Snow, Inc., and Marketing S.R.C.

64 Kempe, E., et al. 1994. *The Quality of Maternal and Neonatal Services in Yemen: Seen Through Women's Eyes*. Stockholm: Save the Children Sweden.

65 UNFPA 2004b, p. 46.

66 JHPIEGO, Save the Children, and Family Care International. 2003. *Shaping Policy for Maternal and Newborn Health: A Compendium of Case Studies*. Baltimore, Maryland: Maternal and Neonatal Health Project, JHPIEGO; and Gay, J., et al., 2003.

67 Mercer, J. 2000. "Family-Centered Maternity Care in Moldova." In: *MotherCare's Initiatives: Actions and Results of 31 Projects: 1993-2000*, edited by S. Jessop, et al. Arlington, Virginia: John Snow International.

68 Glatleider, P., P. Paluzzi, and C. Conroy. 2000. "Changing the Way Maternity Care Is Delivered in the Ukraine." In: *MotherCare's Initiatives: Actions and Results of 31 Project:, 1993-2000*, edited by S. Jessop, et al. Arlington, Virginia: John Snow International.

69 MotherCare/SEATS. 2000. "MotherCare/SEATS, JSI Collaborative Project in Novosibirsk and Primorksy Krai, Russia." In: *MotherCare's Initiatives: Actions and Results of 31 Projects: 1993-2000*, edited by S. Jessop, et al. Arlington, Virginia: John Snow International.

70 Campero, L., et al. 1998. "'Alone I Wouldn't Have Known What To Do': A Qualitative Study on Social Support During Labor and Delivery in Mexico." *Social Science and Medicine* 47(3): 395-403; and Langer, A., et al. 1993. "The Latin American Trial of Psychosocial Support During Pregnancy: A Social Intervention Evaluated Through an Experimental Design." *Social Science and Medicine* 36(4): 495-507.

71 Carter, M. W. 2002. "'Because He Loves Me': Husbands' Involvement in Maternal Health in Rural Guatemala." *Culture, Health, and Sexuality* 4(3): 259-279

72 Abdel-Tawab, N., et al. 2002. "Recovery from Abortion and Miscarriage in Egypt: Does Counseling Husbands Help?" Ch. 10 in: *Responding to Cairo: Case Studies of Changing Practice in Reproductive Health and Family Planning*, edited by N. Haberland and D. Measham. 2002. New York: The Population Council.

73 Program for Appropriate Technology in Health. 2002. "Men and Reproductive Health Programme Examples: India: Nandesari." Reproductive Health Outlook web site: www.rho.org/html/menrh_progexamples.htm#india-nandesari, last accessed 3 May 2004; and Raju, S., and A. Leonard. 2000. *Men as Supportive Partners in Reproductive Health: Moving from Rhetoric to Reality*, pp. 46-47, 52. New Delhi: South and East Asia Regional Office, the Population Council.

74 UNFPA. 2003. *Maternal Mortality Update 2002: A Focus on Emergency Obstetric Care*, pp. 17-34. New York: UNFPA.

75 UNFPA 2004a.

76 UNFPA 2004b, p. 46; and UNFPA 2004a.

CHAPTER 8

1 UNAIDS. 2004. *2004 Report on the Global AIDS Epidemic: 4th Global Report* (UNAIDS/04.16E). Geneva: UNAIDS.

2 United Nations. 1995. *Population and Development*, vol. 1: *Programme of Action adopted at the International Conference on Population and Development: Cairo: 5-13 September 1994*, paragraph 8.28. New York: Department of Economic and Social Information and Policy Analysis, United Nations.

3 Ibid., paragraph 8.29.

4 Ibid., paragraph 7.28.

5 Askew, I., and M. Berer. 2003. "The Contribution of Sexual and Reproductive Health Services to Fight Against HIV/AIDS: A Review." *Reproductive Health Matters* 11(22): 51-73.

6 Berkley, S., P. Piot, and A. Whiteside. 2003. "Scenarios: The Global Health Crisis," p. 1. Produced for the World Economic Forum Annual Meeting, Davos, Switzerland, 23-28 January 2003.

7 Bell, C., S. Devarajan, and H. Gersbach. 2003. *The Long-run Economic Cost of AIDS: Theory and an Application to South Africa*. Policy Research Working Paper. No. 155. Washington, D.C.: The World Bank.

8   Global HIV Prevention Working Group. 2003. *Access to HIV Prevention: Closing the Gap*. Web site: http://www.kff.org/hivaids/loader.cfm?url=/commonspot/security/getfile.cfm&PageID=14225, last accessed 30 June 2004.

9   Ibid.

10 Askew and Berer 2003, p. 57.

11 See the June 2002 special issue on integration: *International Family Planning Perspectives* 28(2).

12 Richey, L. A. 2003. "HIV/AIDS in the Shadows of Reproductive Health Interventions," p. 31. *Reproductive Health Matters* 11(22): 30-35.

13 International Planned Parenthood Federation Western Hemisphere Region. 2000. *Opening Windows to Gender: A Case Study of a Major International Population Agency.* IPPF/WHR Working Paper. No. 1. London: International Planned Parenthood Federation Western Hemisphere Region; Becker, J., and E. Leitman. 1997. "Introducing Sexuality within Family Planning: Three Positive Experiences from Latin America and the Caribbean." *Quality/Calidad/Qualité.* No. 8. New York: The Population Council; and Hardee, K. 2004. "The Intersection of Access, Quality of Care and Gender in Reproductive Health and STI/HIV Services: Evidence from Kenya, India and Guatemala." Draft. Washington, D.C.: Interagency Gender Working Group and the POLICY Project, the Futures Group.

14 Banda, H. N., S. Bradley, and K. Hardee. 2004. *Provision and Use of Family Planning in the Context of HIV/AIDS in Zambia: Perspectives of Providers, Family Planning and Antenatal Care Clients and HIV-Positive Women.* Final report. Washington, D.C.: POLICY Project, the Futures Group, and the Support for Analysis and Research (SARA) Project; and Gichuhi, W., and S. Bradley. 2004. *Provision and Use of Family Planning in the Context of HIV/AIDS in Kenya: Perspectives of Providers, Family Planning and Antenatal Care Clients, and HIV-Positive Women.* Final report. Washington, D.C.: The Futures Group, POLICY Project and SARA Project, Academy for Educational Development.

15 Rothenberg, R. B., et al. 2000. "The Effect of Treating Sexually Transmitted Diseases on the Transmission of HIV in Dually Infected Persons: A Clinic-based Estimate: Ad Hoc STD/HIV Transmission Group." *Sexually Transmitted Diseases* 27(7): 411-416. Cited in: Askew and Berer 2003.

16 Grosskurth, H., et al. 1995. "Impact of Improved Treatment of Sexually Transmitted Diseases on HIV Infection in Rural Tanzania:

Randomised Controlled Trial." *The Lancet.* 346(8974): 530-536.

17 Sloan, N. L., et al. 2000. "Screening and Syndromic Approaches to Identify Gonorrhea and Chlamydial Infection Among Women." *Studies in Family Planning* 31(1): 55-68.

18 Lush, L., G. Walt, and J. Ogden. 2003. "Transferring Policies for Treating Sexually Transmitted Infections: What's Wrong with Global Guidelines?" *Health Policy and Planning* 18(1): 18-30.

19 Askew and Berer 2003, pp. 53-54.

20 Gray, R. H. 2003. "Guest Commentary: Contraception and HIV Acquisition in Rakai, Uganda." *The Pop Reporter* 3(7). Baltimore, Maryland: INFO Project, Center for Communication Programs, Johns Hopkins Bloomberg School of Public Health.

21 Chaya, N., K-H. Amen, and M. Fox. 2002. *Condoms Count: Meeting the Need in the Era of HIV/AIDS: The PAI Report Card 2002.* Washington, D.C.: Population Action International.

22 WHO. 2002. *WHO Information Update regarding Reuse of the Female Condom.* Geneva: WHO. Web site: www.who.int/reproductive-health/rtis/reuse.en.html, accessed 20 February 2004.

23 Rivers, K., et al. 1998. "Gender Relations, Sexual Communication and the Female Condom." *Critical Public Health* 8(4): 273-290. Cited in: "Programming for the Female Condom: Why is the Jury Still Out," by T. Hatzell. 2001. Presentation at the Forum, Technical Update on the Female Condom, Washington, D.C., 18 December 2001.

24 Finger, W. R. 2000. "Zimbabwe Project Promotes Female Condom Use." *Network* 20(2): 20.

25 Camlin, C., and C. Chimbwete. 2003. "Does Knowing Someone with AIDS Affect Condom Use: An Analysis from South Africa." *AIDS Education and Prevention* 15(3): 231-244.

26 Brady, M. 2003. "Preventing Sexually Transmitted Infections and Unintended Pregnancy, and Safeguarding Fertility: Triple Protection Needs of Young Women," p. 137. *Reproductive Health Matters* 11(22): 134-141.

27 The Microbicide Initiative. 2002. *Mobilization for Microbicides: The Decision Decade.* New York: The Rockefeller Foundation. Cited in: Chaya, Amen, and Fox 2002.

28 International Planned Parenthood Federation South Asia Regional Office and UNFPA. 2004. *Integrating HIV Voluntary Counselling and Testing Services into Reproductive Health

Settings: Stepwise Guidelines for Programme Planners, Managers and Service Providers.* London and New York: International Planned Parenthood Federation South Asia Regional Office and UNFPA.

29 "The Glion Call to Action on Family Planning and HIV/AIDS in Women and Children: 3-5 May 2004." Geneva: WHO.

30 WHO and UNAIDS. 2003. *Treating 3 Million by 2005: Making it Happen: The WHO Strategy.* Geneva: WHO and UNAIDS.

31 Ibid., p. 31.

32 Banda, Bradley, and Hardee 2004.

33 Pisani, E., et al. 2003. "Back to Basics in HIV Prevention: Focus on Exposure." *British Medical Journal* 326(7403):1384-1387.

34 Banda, Bradley, and Hardee 2004.

35 Askew and Berer 2003, p. 55.

36 Office of Health, Ministry of Health and Prevention, Government of Senegal. 2002. *Mapping Integration of FP/MCH and STI/HIV/AIDS Services in Senegal's Kaolack Region,* p. 40. Arlington, Virginia: Advance Africa.

37 Hardee, K., and J. Smith. 2000. "Increasing Returns on Reproductive Health Services in the Era of Health Sector Reform." POLICY Occasional Paper. No. 5. Washington D.C.: The Futures Group International.

38 Berer, M. 2003. "HIV/AIDS, Sexual and Reproductive Health: Intimately Related," p. 9. *Reproductive Health Matters* 11(22): 6-11.

39 Scillia, A. September-October 2001. "Senegal: Ray of Hope as Transmission Rates Slow." *The Courier ACP-EU,* pp. 16-17; and UN Office of the Coordination of Humanitarian Affairs. n.d. "Senegal: Country Profile." IRIN PlusNews. Web site: www.irinnews.org/AIDS/senegal.asp, accessed 23 February 2004.

40 Chase, M. 25 February 2004. "Saying No to 'Sugar Daddies': Can a Financial Prophylactic Shield Girls From Liaisons That Spread AIDS in Africa?" *The Wall Street Journal,* pp. B1-2.

41 UNFPA. 2004. *Investing in People: National Progress in Implementing the ICPD Programme of Action,* p. 128. New York: UNFPA.

42 Ibid., p. 130.

CHAPTER 9

1   United Nations. 1995. *Population and Development,* vol. 1: *Programme of Action adopted at the International Conference on Population and

Development: Cairo: 5-13 September 1994,* paragraph 7.44. New York: Department of Economic and Social Information and Policy Analysis, United Nations.

2   Ibid., paragraphs 7.45-7.48.

3   Price, N. 2004. "Addressing the Reproductive Health Needs and Rights of Young People since ICPD: The Contribution of UNFPA and IPPF: Synthesis Report." Unpublished report. London; Söborg, Denmark; and Heidelberg: Options, Euro Health Group, University of Heidelberg.

4   United Nations. 1999. *Key Actions for the Further Implementation of the Programme of Action of the International Conference on Population and Development* (A/S-21/5/Add.1), paragraph 70. New York: United Nations.

5   UNFPA. 2004. *Investing in People: National Progress in Implementing the ICPD Programme of Action,* p. 60. New York: UNFPA.

6   UNFPA. 9 May 2003. "UNFPA Global Population Policy Update," Issue 3. New York: UNFPA; and UNFPA. 21 April 2003. "UNFPA Global Population Policy Update," Issue 1. New York: UNFPA

7   UNFPA. 5 April 2004. "UNFPA Global Population Policy Update," Issue 5. New York: UNFPA; and UNFPA. 18 July 2004. "UNFPA Global Population Policy Update," Issue 6. New York: UNFPA

8   UNFPA. 2003a. "UNFPA and Young People: Imagine." New York: UNFPA.

9   Ibid.

10 UNFPA. 2003b. "Fast Facts on Adolescents and Youth." Web site: www.unfpa.org/adolescents/facts.htm, last accessed 16 February 2004.

11 United Nations. 2004. *Review and Appraisal of the Progress made in Achieving the Goals and Objectives of the Programme of Action of the International Conference on Population and Development: Report of the Secretary-General* (E/CN.9/2004/3). New York: United Nations.

12 United Nations. 2003. *World Population Prospects: The 2002 Revision.* New York: Population Division, Department Economic and Social Affairs, United Nations.

13 United Nations 2004.

14 UNESCO. 2002. *EFA Global Monitoring Report 2002: Education for All: Is the World on Track?* Paris: UNESCO. Web site: www.unesco.org, last accessed 2 February 2004.

15 United Nations 2004.

16 United Nations 1995, paragraph 7.46.

17 Curtain, R. 2003. "The Case for Investing More in Young People as an Integral part of a Poverty Eradication Strategy." Unpublished paper. Melbourne, Australia: Curtain Consulting.

18 A Mother's Promise. n.d. "Keeping Young People Healthy." Fact sheet. Washington, D.C.: A Mother's Promise. Web site: www.ppfa.org/promise/learn_more.asp, last accessed 13 May 2004.

19 UNFPA. 2003c. *The State of World Population 2003: Making 1 Billion Count: Investing in Adolescents' Health and Rights.* New York: UNFPA.

20 UNFPA 2003a.

21 UNICEF. 2001. *Early Marriage: Child Spouses.* Innocenti Digest. No. 7. Florence, Italy: UNICEF. Innocenti Research Centre. Web site: www.unicef-icdc.org/publications/pdf/digest7e.pdf, last accessed 29 June 2004.

22 Ibid.

23 United Nations 2004.

24 United Nations. 2002. *World Population Monitoring 2002: Reproductive Rights and Reproductive Health: Selected Aspects* (ESA/P/WP.717). New York: United Nations.

25 UNFPA 2003a.

26 Haberland, N., and D. Measham (eds). 2002. *Responding to Cairo: Case Studies of Changing Practice in Reproductive Health and Family Planning.* New York: The Population Council

27 UNFPA 2003a.

28 UNFPA. 20 April 2004. "Youth Leaders Advise UNFPA on their Generation's Rights and Needs." Press release. New York: UNFPA.

29 Price 2004.

CHAPTER 10

1 WHO, UNFPA, and UNHCR. 1999. *Reproductive Health in Refugee Situations: An Inter-Agency Field Manual.* Geneva: UNHCR.

2 After a mother's death, surviving children are three to ten times more likely to die within two years than children who live with both parents. See: UNICEF. 2001. *The State of the World's Children 2001: Early Childhood.* New York: UNICEF.

3 Krause, S., et al. 2001. *Assessment of Reproductive Health for Refugees in Zambia.* New York: Women's Commission for Refugee Women and Children.

4 UNAIDS and WHO. 2003. *AIDS Epidemic Update: December 2003.* Geneva: UNAIDS

5 PROFAMILIA. 2001. "Sexual and Reproductive Health in Underserved Conditions: A Survey of the Situation of Displaced Women in Colombia." Bogotá, Colombia: PROFAMILIA.

6 Rehn, E., and E. Johnson Sirleaf. 2002. *Women, War and Peace: The Independent Experts' Assessment on the Impact of Armed Conflict on Women and Women's Role in Peacebuilding: Progress of the World's Women 2002,* vol.2. New York: UNIFEM.

7 UNFPA. 2001. *Population Issues Briefing Kit 2001,* p. 17. New York: UNFPA.

CHAPTER 11

1 The Statement of Commitment may be found at: www.unfpa.org/ipci/index.htm.

2 See web site: www.south-south-ppd.org/, accessed 4 June 2004.

3 Agenda 21 of the 1992 UN Conference on the Environment and Development had included some nominal costs but governments did not agree on them and they were presented as provisional.

4 United Nations. 1995. *Population and Development,* vol. 1: *Programme of Action adopted at the International Conference on Population and Development: Cairo: 5-13 September 1994,* paragraphs 13.14-13.16. New York: Department of Economic and Social Information and Policy Analysis, United Nations.

5 This estimate did not take appropriate account of family planning (and some other reproductive health) supplies or their delivery costs. These costs and other components of reproductive health infrastructure were included in an aggregate total addressing infrastructure for service delivery.

6 The report of the methodology can be found in: Schwartlander, B., et al. 2001. "Resource Needs for HIV/AIDS." *Science* 292(5526): 2434-2436. See also: Stover, J., et al. 2002. "Can We Reverse the HIV/AIDS Pandemic with an Expanded Response?" *The Lancet* 360(9326): 73-77.

7 The estimate for the ICPD components of mass media, education programmes, and additional condoms for HIV/AIDS prevention was $1.7 billion; some $200 million higher than projected in 1994.

8 A technical consultation on integrating reproductive health and HIV/AIDS programming was held in May 2004, hosted by UNFPA. A report of the meeting and the background papers will be available on the web site: www.unfpa.org.

9 New vertical programmes may also divert resources and skilled personnel from existing health system efforts. Anecdotal evidence about the negative impact of large resource flows to HIV/AIDS is easy to find. Serious studies of the policy and programme impacts are in their early stages.

10 The increased use of sector-wide funding and integrated programming makes it difficult to track resource streams for particular activities. Estimating the proportion of a multi-purpose allocation going to particular elements is inherently difficult. Evidence shows that support for family planning has been a declining share of total allocations. (See United Nations. 2004. *Flow of Financial Resources for Assisting in the Implementation of the Programme of Action of the International Conference on Population and Development: A 10-year Review: Report of the Secretary-General* (E/CN.9/2004/4). New York: United Nations.

11 Caution should be also exercised in interpreting the high level of reported domestic expenditures even in the countries that do report. The Programme of Action service delivery estimates were made for primary health care level interventions and, in the area of HIV/AIDS, only for selected prevention activities. Reports of domestic expenditures use the general functional categories of the basic costed package, but higher level service provision (including hospital care and lower-level care given at these more expensive facilities) are frequently included. Since 1999, both donor and domestic HIV/AIDS reports are not restricted to only the ICPD-specified prevention activities.

12 The project is directed by Jeffrey Sachs, Special Adviser to the Secretary-General, and UN Development Programme Administrator Mark Malloch Brown, and is supported by Task Forces of international experts. They are analysing priority interventions to accelerate progress and working to ensure that Poverty Reduction Strategies in developing countries give priority to attaining the Millennium Development Goals. Details can be found at: www.unmillenniumproject.org.

13 The Interim Reports of the Task Forces on Child and Maternal Mortality and on Primary Education and Gender Equality both endorse universal access to reproductive health services (an ICPD goal) as fundamental to the attainment of the Millennium Development Goals (accessible at the project website, see the previous note).

14 A few countries, mostly Nordic countries, have attained this level of support. In contrast, the United States provides the lowest share of relative to GDP of major donor country.

Sources for Boxes

CHAPTER 1

1 UNFPA. 2004. *Working from Within: Culturally Sensitive Approaches in UNFPA Programming.* New York. UNFPA.

2 United Nations. 2003. *World Population Prospects: The 2002 Revision.* New York: Population Division, Department Economic and Social Affairs, United Nations; and United Nations. 2003. *The Impact of AIDS* (ESA/P/WP.185). New York: Population Division, Department of Economic and Social Affairs, United Nations.

CHAPTER 2

3 United Nations. 1995. *Population and Development,* vol. 1: *Programme of Action adopted at the International Conference on Population and Development: Cairo: 5-13 September 1994.* New York: Department of Economic and Social Information and Policy Analysis, United Nations; and United Nations. 1999. *Key Actions for the Further Implementation of the Programme of Action of the International Conference on Population and Development* (A/S-21/5/Add.1). New York: United Nations.

4 WHO and the World Bank. 2002. *Dying for Change: Poor People's Experience of Health and Ill Health.* The Voices of the Poor Study. Geneva: WHO; Gwatkin, D., et al. 2003. *Initial Country-level Information about Socio-Economic Differences in Health, Nutrition and Population,* 2nd edition. Washington, D.C.: The World Bank; UNFPA. 2002. *The State of World Population 2002: People, Poverty and Possibilities: Making Development Work for the Poor.* New York: UNFPA; and United Nations. 2004. *Review and Appraisal of the Progress Made In Achieving the Goals and Objectives of the Programme of Action of the International Conference on Population and Development: Report of the Secretary-General* (E/CN.9/2004/3). New York: Commission on Population and Development, Economic and Social Council, United Nations.

CHAPTER 3

5 French, H. 2001. "Greening Globalization." *Georgetown Journal of International Affairs* 2(1): 25; French, H. 2004. "Linking Globalization, Consumption, and Governance," p. 145. Ch. 7 in: *State of the World 2004,* by the Worldwatch Institute. 2004. New York: W.W. Norton; FAO. 2003a. *State of the World's Forest,* p. 1. Rome: FAO; FAO. 2004. *Fisheries Commodities Production and Trade 1976-2000.* Electronic database. Web site: www.fao.org/fi/statist/fisoft/FISHPLUS.asp, last accessed 24 February 2004; and FAO. 2003b. *World Agriculture:*

Towards 2015/2030, p.197. Rome: FAO; Myers, R., and B. Worm. 2003. "Rapid Worldwide Depletion of Predatory Fish Communities." *Nature* 423: 280-283; Revkin, A. 15 May 2003. "Commercial Fleets Reduced Big Fish by 90 Percent, Study Says." *The New York Times*, p. A16; and SeaWeb. 2003. "Cover Study of *Nature* Provides Startling New Evidence that Only 10 Percent of All Large Fish are Left in the Ocean." Press release. Washington, D.C.: SeaWeb; Myers, R., and B. Worm. 2003. "Rapid Worldwide Depletion of Predatory Fish Communities." *Nature* 423: 280-283; Revkin, A. 15 May 2003. "Commercial Fleets Reduced Big Fish by 90 Percent, Study Says." *The New York Times*, p. A16; and SeaWeb. 2003. "Cover Study of *Nature* Provides Startling New Evidence that Only 10 Percent of All Large Fish are Left in the Ocean." Press release. Washington, D.C.: SeaWeb; and Engelman, R., et al. 2000. *People in the Balance: Population and Natural Resources at the Turn of the Millennium.* Washington, D.C.: Population Action International; and Gardner, G., E. Assadourian, and R. Sarin. 2004. "The State of Consumption Today," p. 17. Ch. 1 in: *State of the World 2004*, by the Worldwatch Institute. 2004. New York: W. W. Norton.

6 Bradsher, K. 18 November 2003. "China Set to Act on Fuel Economy." *The New York Times*, p. A1; and French, H. 2001. "Greening Globalization." *Georgetown Journal of International Affairs* 2(1): 25.

CHAPTER 4

8 Balk, D., et al. 2004. "Methodologies to Improve Global Population Estimates in Urban and Rural Areas." Paper presented at the annual meeting of the Population Association of America, Boston, Massachusetts, 2 April, 2004. Extended abstract available on web site: http://paa2004.princeton.edu/abstractViewer.asp?submissionId=41835, last accessed 19 May 2004; Balk, D., et al. 2003. *Spatial Analysis of Childhood Mortality in West Africa.* DHS Geographic Series 1. Calverton, Maryland: ORC Macro and Center for International Earth Science Information Network (CIESIN), Columbia University; and Hugo, G., A. Champion, and A. Lattes. 2003. "Toward a New Conceptualization of Settlements for Demography." *Population and Development Review* 29(2): 277-297.

CHAPTER 5

9 UNFPA and Italian Association for Women and Development (AIDOS). n.d. "Final Summary Report: The UNFPA Conference on Gender-Based Violence," Rome, Italy, 15-19 September 2003. New York:

Technical Support Division, UNFPA; and Donnay, F., Chief, Reproductive Health Branch, Technical Support Division, UNFPA. 2003. Personal communication.

10 WHO. 2003. *'En-gendering' the Millennium Development Goals (MDGs) on Health.* Geneva: Department of Gender and Women's Health, WHO.

11 Grown, C., G. R. Gupta, and Z. Khan. 2003. "Background Paper of the Task Force on Education and Gender Equality: Promises to Keep: Achieving Gender Equality and the Empowerment of Women." New York: UN Millennium Project; and McPhedran, M., et al. 2000. "The First CEDAW Impact Study: Final Report." Toronto: Centre for Feminist Research, York University, and the International Women's Rights Project.

CHAPTER 6

12 WHO. 2003. *Reproductive Health: Draft Strategy to Accelerate Progress towards the Attainment of International Development Goals and Targets* (EB113/15 Add.1). Geneva: WHO.

13 Singh, S., et al. 2004. *Adding It Up: The Benefits of Investing in Sexual and Reproductive Health Care.* New York: The Alan Guttmacher Institute and UNFPA.

14 Data provided by the UNFPA Strategic Planning Office, incorporating responses to the annual Multi-year Funding Framework (MYFF) survey.

15 Kols, A. J., and J. E. Sherman. 1998. "Family Planning Programs: Improving Quality." *Population Reports.* Series J. No. 47. Baltimore, Maryland: Population Information Program, Johns Hopkins University School of Public Health, Population Information Program

16 POLICY Project. 2000. *Health Reform, Decentralization and Participation in Latin America: Protecting Sexual and Reproductive Health.* Washington, D.C.: POLICY Project, the Futures Group. Also see: UNFPA. 1999. "Partnership and Empowerment." Ch. 4 in: UNFPA. 1999. *The State of World Population 1999: 6 Billion: A Time for Choices.* New York: UNFPA and the references cited therein; and Huezo, C. 2003. "Strengthening the Quality of Reproductive Health Care: IPPF's Quality Improvement Programme." *IPPF Medical Bulletin* 37(5): 1-3.

17 UNFPA. 2001. *Reproductive Health Commodity Security: Partnerships for Change: A Global Call to Action.* New York: UNFPA. Web site: www.unfpa.org/upload/lib_pub_file/135_filename_rhcstrategy.pdf, last accessed 17 June 2004.

CHAPTER 7

18 UNFPA and EngenderHealth. 2003. *Obstetric Fistula: Needs Assessment Report: Findings from Nine African Countries.* New York: UNFPA and EngenderHealth. Web site: www.unfpa.org/fistula/docs/fistula-needs-assessment.pdf, last accessed 27 May 2004.

19 Freedman, L., et al. 2004. "Interim Report of Task Force 4 on Child Health and Maternal Health." New York: UN Millennium Project; and Wardlaw, T., and D. Maine. 1999. "Process Indicators for Maternal Mortality Programs." In: *Safe Motherhood Initiatives: Critical Issues,* edited by M. Berer and T. K. S. Ravindran. 1999. Oxford, United Kingdom: Blackwell Science for Reproductive Health Matters.

20 UNFPA. 5 February 2004. "The New Route to Safer Childbirth in Rural Senegal." News Feature. Web site: www.unfpa.org/news/news.cfm?ID=389&Language=1, last accessed 17 June 2004.

21 Santillan, D., and M. E. Figueroa. 2001. *Implementing a Client Feedback System to Improve NGO Healthcare Services in Peru.* QA Operations Research Results Series. No. 2(2). Bethesda, MD: Quality Assurance Project for the US Agency for International Development (USAID).

22 UNFPA. 2004. *Saving Mothers' Lives: The Challenge Continues.* Brochure. New York: UNFPA.

CHAPTER 8

23 UNAIDS, UNFPA, and UNIFEM. 2004. *Women and HIV/AIDS: Confronting the Crisis: A Joint Report by UNAIDS, UNFPA and UNIFEM.* Geneva and New York: UNAIDS, UNFPA, and UNIFEM. Also see web site: http://womenandaids.unaids.org/default.html, last accessed 17 June 2004.

24 The Population Council and UNFPA. 2002. *HIV/AIDS Prevention Guidance for Reproductive Health Professionals in Developing-Country Setting,* p. 18. New York: The Population Council and UNFPA.

25 International Planned Parenthood Federation and UNFPA. 2004. *Integrating HIV Voluntary Counselling and Testing into Reproductive Health Services. Stepwise Guidelines for Programme Planners, Managers and Service Providers.* New York: UNFPA and London: International Planned Parenthood Federation South Asia Regional Office.

26 The Alan Guttmacher Institute. 2003. *A, B and C in Uganda: Roles of Abstinence, Monogamy and Condom Use in HIV Decline: Executive Summary.* New York: The Alan Guttmacher Institute.

27 Feldman, R., and C. Maposhere. 2003. "Safer Sex and Reproductive Choice: Findings From 'Positive Women: Voices and Choices' in Zimbabwe." *Reproductive Health Matters* 11(22): 162-173.

CHAPTER 9

29 UNFPA. 19 April 2004. "Multi-Media Centre Provides Hands on Training for Youth in Benin." Web site: www.unfpa.org/news/news.cfm?ID=444&Language=1, last accessed 17 June 2004.

30 UNFPA. 4 June 2004. "Global Health Council Child Marriage: Advancing the Global Agenda." Statement by Thoraya Ahmed Obaid, Executive Director, UNFPA. New York: UNFPA; UNFPA. 4 June 2004. "Married Adolescents Ignored in Global Agenda, Says UNFPA." Press Release. New York: UNFPA; and UNFPA. 2004. "Too Brief a Child: Voices of Married Adolescents." Video produced for UNFPA by Spark Media. New York: UNFPA.

31 African Youth Alliance. 2003. Country Profiles 2003 for Botswana, Ghana, Uganda, and the Republic of Tanzania. New York: African Youth Alliance.

32 UNFPA, WHO, and UNICEF. 2003. *Adolescents: Profiles in Empowerment.* New York: UNICEF.

CHAPTER 10

33 UNFPA. 2001. *The Impact of Conflict on Women and Girls.* New York: UNFPA.

34 Inter-Agency Working Group on Reproductive Health in Refugee Settings. Forthcoming. *Reproductive Health in Refugee and IDP Situations: Evaluation Report.* Geneva: Inter-Agency Working Group on Reproductive Health in Refugee Settings.

CHAPTER 11

35 UNFPA. 2004. "El caso de Nicaragua: 1998-2003." PowerPoint presentation at the Regional Planning Meeting, Latin America and Caribbean Division, UNFPA, New York, New York, 19-24 January 2004.

Monitoring ICPD Goals – Selected Indicators

	Indicators of Mortality			Indicators of Education				Reproductive Health Indicators			
	Infant mortality Total per 1,000 live births	Life expectancy M/F	Maternal mortality ratio	Primary enrolment (gross) M/F	Proportion reaching grade 5 M/F	Secondary enrolment (gross) M/F	% Illiterate (>15 years) M/F	Births per 1,000 women aged 15-19	Contraceptive Prevalence Any method	Contraceptive Prevalence Modern methods	HIV prevalence rate (%) (15-49) M/F
World Total	56	63.3 / 67.6						50	61	54	
More developed regions (*)	8	72.1 / 79.4						27	69	55	
Less developed regions (+)	61	61.7 / 65.1						53	59	54	
Least developed countries (‡)	97	48.8 / 50.5						124			
AFRICA (1)	89	47.9 / 50.0						107	27	20	
EASTERN AFRICA	97	42.4 / 43.8						117	22	17	
Burundi	107	40.4 / 41.4	1,000	80 / 62	68 / 59	12 / 9	42 / 56	50	16	10	5.2 / 6.8
Eritrea	73	51.2 / 54.2	630	67 / 54		33 / 22		115	8	5	2.3 / 3.0
Ethiopia	100	44.6 / 46.3	850	75 / 53	63 / 59	23 / 15	51 / 66	100	8	6	3.8 / 5.0
Kenya	69	43.5 / 45.6	1,000	97 / 95		34 / 30	10 / 21	78	39	32	4.6 / 8.9
Madagascar	91	52.5 / 54.8	550	106 / 102	33 / 34			137	19	12	1.4 / 1.9
Malawi	115	37.3 / 37.7	1,800	149 / 143	61 / 47	39 / 29	24 / 51	163	31	26	12.4 / 16.0
Mauritius (2)	16	68.4 / 75.8	24	106 / 106	99 / 99	81 / 78	12 / 19	33	75	49	
Mozambique	122	36.6 / 39.6	1,000	110 / 87	56 / 47	16 / 10	38 / 69	105	6	5	10.6 / 13.8
Rwanda	112	38.8 / 39.7	1,400	118 / 116	39 / 41	15 / 14	25 / 37	50	13	4	4.4 / 5.7
Somalia	118	46.4 / 49.5	1,100					213			
Uganda	86	45.4 / 46.9	880	139 / 134		19 / 15	21 / 41	211	23	18	3.7 / 4.9
United Republic of Tanzania	100	42.5 / 44.1	1,500	70 / 69	76 / 80		15 / 31	120	25	17	7.6 / 9.9
Zambia	105	32.7 / 32.1	750	81 / 76	79 / 75	27 / 21	14 / 26	145	34	23	14.1 / 18.9
Zimbabwe	58	33.7 / 32.6	1,100	100 / 98		45 / 40	6 / 14	92	54	50	21.0 / 28.4
MIDDLE AFRICA (3)	116	41.6 / 43.8						200	23	5	
Angola	140	38.8 / 41.5	1,700			21 / 17		220	6	5	3.4 / 4.4
Cameroon	88	45.1 / 47.4	730	115 / 99		36 / 29	23 / 40	121	19	7	6.0 / 7.9
Central African Republic	100	38.5 / 40.6	1,100	79 / 53			35 / 67	132	28	7	11.9 / 15.1
Chad	115	43.7 / 45.7	1,100	90 / 57	58 / 48	17 / 5	46 / 63	195	8	2	4.2 / 5.4
Congo, Democratic Republic of the (4)	120	40.8 / 42.8	990					230	31	4	3.7 / 4.8
Congo, Republic of	84	46.6 / 49.7	510	88 / 83		37 / 27	11 / 23	146			4.3 / 5.6
Gabon	57	55.8 / 57.5	420	135 / 134				113	33	12	7.1 / 9.1
NORTHERN AFRICA (5)	49	64.5 / 68.2						36	47	42	
Algeria	44	68.1 / 71.3	140	112 / 104	95 / 97	69 / 74	22 / 40	16	64	50	0.1 / <0.1
Egypt	41	66.7 / 71.0	84	100 / 94	99 / 99	91 / 85	33 / 56	47	56	54	0.1 / <0.1
Libyan Arab Jamahiriya	21	70.8 / 75.4	97	114 / 114		102 / 108	8 / 29	7	40	26	
Morocco	42	66.8 / 70.5	220	113 / 101	84 / 83	45 / 36	37 / 62	25	50	42	
Sudan	77	54.1 / 57.1	590	63 / 54		34 / 30	29 / 51	55	8	7	1.9 / 2.6
Tunisia	23	70.8 / 74.9	120	114 / 109	95 / 96	78 / 81	17 / 37	7	60	51	<0.1 / <0.1
SOUTHERN AFRICA	52	43.9 / 49.1						66	53	51	
Botswana	57	38.9 / 40.5	100	103 / 103	87 / 92	70 / 75	24 / 18	91	40	39	31.7 / 43.1
Lesotho	92	32.3 / 37.7	550	123 / 125	60 / 74	30 / 38	26 / 10	53	30	30	25.4 / 32.4
Namibia	60	42.9 / 45.6	300	106 / 106	94 / 94	57 / 65	16 / 17	78	29	26	18.4 / 24.2
South Africa	48	45.1 / 50.7	230	107 / 103	65 / 64	83 / 90	13 / 15	66	56	55	18.1 / 23.5
Swaziland	78	33.3 / 35.4	370	103 / 98	69 / 79	45 / 45	18 / 20	45	28	26	35.7 / 41.7
WESTERN AFRICA (6)	90	49.0 / 50.3						119	15	8	
Benin	93	48.4 / 53.0	850	122 / 86	89 / 78	35 / 16	45 / 74	107	19	7	1.7 / 2.1
Burkina Faso	93	45.2 / 46.2	1,000	51 / 36	68 / 71	12 / 8	82 / 92	136	12	5	3.6 / 4.8
Côte d'Ivoire	101	40.8 / 41.2	690	92 / 68				116	15	7	6.0 / 8.1
Gambia	81	52.7 / 55.5	540	82 / 75		40 / 28		125	10	9	1.0 / 1.3

	Indicators of Mortality			Indicators of Education				Reproductive Health Indicators			
	Infant mortality Total per 1,000 live births	Life expectancy M/F	Maternal mortality ratio	Primary enrolment (gross) M/F	Proportion reaching grade 5 M/F	Secondary enrolment (gross) M/F	% Illiterate (>15 years) M/F	Births per 1,000 women aged 15-19	Contraceptive Prevalence — Any method	Contraceptive Prevalence — Modern methods	HIV prevalence rate (%) (15-49) M/F
Ghana	58	56.5 / 59.3	540	85 / 78	67 / 65	41 / 34	18 / 34	76	22	13	2.6 / 3.5
Guinea	102	48.8 / 49.5	740	88 / 66	90 / 77			163	6	4	2.7 / 3.7
Guinea-Bissau	120	43.8 / 46.9	1,100					197	8	4	
Liberia	147	40.7 / 42.2	760				28 / 61	227	6	6	5.1 / 6.7
Mali	119	48.0 / 49.1	1,200	65 / 49	88 / 79		73 / 88	191	8	6	1.6 / 2.2
Mauritania	97	50.9 / 54.1	1,000	88 / 85	54 / 56	25 / 19	49 / 69	104	8	5	0.6 / 0.7
Niger	126	45.9 / 46.5	1,600	47 / 32	73 / 68	8 / 5	75 / 91	233	14	4	1.0 / 1.4
Nigeria	79	51.1 / 51.8	800	107 / 86			26 / 41	103	15	9	4.6 / 6.2
Senegal	61	50.8 / 55.1	690	79 / 72	70 / 65	22 / 15	51 / 70	86	13	8	0.7 / 0.9
Sierra Leone	177	33.1 / 35.5	2,000	93 / 65		31 / 22		212	4	4	
Togo	81	48.2 / 51.1	570	136 / 112	88 / 80		26 / 55	81	26	9	3.6 / 4.7
ASIA	**53**	**65.5 / 69.0**						**35**	**64**	**58**	
EASTERN ASIA (7)	34	69.7 / 74.7						5	82	81	
China	37	68.9 / 73.3	56	114 / 114			5 / 13	5	84	83	0.2 / 0.1
Democratic People's Republic of Korea	45	60.5 / 66.0	67					2	62	53	
Hong Kong SAR, China (8)	4	77.3 / 82.8		108 / 108		78 / 78		6	86	80	0.1 / 0.1
Japan	3	77.9 / 85.1	10 [9]	101 / 101		102 / 103		4	56	51	<0.1 / <0.1
Mongolia	58	61.9 / 65.9	110	97 / 100	87 / 90	69 / 83	2 / 3	54	67	54	<0.1 / <0.1
Republic of Korea	5	71.8 / 79.3	20	102 / 102	100 / 100	91 / 91		3	81	67	0.1 / <0.1
SOUTH-EASTERN ASIA	41	64.4 / 69.1						42	57	49	
Cambodia	73	55.2 / 59.5	450	130 / 116	71 / 70	27 / 16	19 / 41	60	24	19	3.7 / 1.6
Indonesia	42	64.8 / 68.8	230	112 / 110	87 / 92	58 / 58	8 / 17	55	57	55	0.2 / <0.1
Lao People's Democratic Republic	88	53.3 / 55.8	650	123 / 106	62 / 63	47 / 34	23 / 45	91	32	29	0.1 / <0.1
Malaysia	10	70.8 / 75.7	41	95 / 95	98 / 96	66 / 73	8 / 15	18	55	30	0.7 / 0.1
Myanmar	83	54.6 / 60.2	360	90 / 90	59 / 61	41 / 38	11 / 19	24	33	28	1.6 / 0.7
Philippines	29	68.0 / 72.0	200	113 / 111	76 / 83	78 / 86	7 / 7	38	47	28	<0.1 / <0.1
Singapore	3	75.9 / 80.3	30				3 / 11	6	62	53	0.4 / 0.1
Thailand	20	65.3 / 73.5	44	100 / 96		85 / 81	5 / 9	49	72	70	2.0 / 1.1
Viet Nam	34	66.9 / 71.6	130	107 / 100	90 / 88	72 / 67	6 / 13	21	78	57	0.7 / 0.3
SOUTH CENTRAL ASIA	68	62.5 / 63.9						54	48	41	
Afghanistan	162	43.0 / 43.3	1,900	44 / -		24 / -		111	5	4	
Bangladesh	64	61.0 / 61.8	380	97 / 98	63 / 68	45 / 49	50 / 69	117	54	43	
Bhutan	54	62.0 / 64.5	420		89 / 93			54	19	19	
India	64	63.2 / 64.6	540	107 / 90	59 / 59	56 / 40		45	48	43	
Iran (Islamic Republic of)	33	68.9 / 71.9	76	94 / 90	94 / 94	79 / 75	17 / 30	33	73	56	0.1 / <0.1
Nepal	71	60.1 / 59.6	740	130 / 113	75 / 81	50 / 37	38 / 74	117	39	35	0.1 / <0.1
Pakistan	87	61.2 / 60.9	500	84 / 62		29 / 19	47 / 71	50	28	20	0.2 / <0.1
Sri Lanka	20	69.9 / 75.9	92	111 / 110			5 / 10	22	66	44	0.0 / <0.1
WESTERN ASIA	44	67.1 / 71.3						47	47	28	
Iraq	83	59.2 / 62.3	250					38	14	10	
Israel	6	77.1 / 81.0	17	114 / 113	100 / 99	95 / 94	3 / 7	17	68	52	
Jordan	24	69.7 / 72.5	41	98 / 99		85 / 87	5 / 14	27	56	39	
Kuwait	11	74.9 / 79.0	5	95 / 94	99 / 98	83 / 88	15 / 19	31	50	41	
Lebanon	17	71.9 / 75.1	150	105 / 101	92 / 96	74 / 81		25	61	37	0.2 / 0.0
Occupied Palestinian Territory	21	70.8 / 74.0	100	104 / 105	97 / 98	82 / 88		94			
Oman	20	71.0 / 74.4	87	84 / 82	96 / 96	79 / 78	18 / 35	66	24	18	0.2 / 0.1

Monitoring ICPD Goals – Selected Indicators

	Indicators of Mortality			Indicators of Education				Reproductive Health Indicators			
	Infant mortality Total per 1,000 live births	Life expectancy M/F	Maternal mortality ratio	Primary enrolment (gross) M/F	Proportion reaching grade 5 M/F	Secondary enrolment (gross) M/F	% Illiterate (>15 years) M/F	Births per 1,000 women aged 15-19	Contraceptive Prevalence Any method	Contraceptive Prevalence Modern methods	HIV prevalence rate (%) (15-49) M/F
Saudi Arabia	21	71.1 / 73.7	23	68 / 66	94 / 94	73 / 65	16 / 31	38	32	29	
Syrian Arab Republic	22	70.6 / 73.1	160	115 / 108	93 / 92	47 / 42	9 / 26	34	36	28	<0.1 / <0.1
Turkey (10)	40	68.0 / 73.2	70	98 / 91		86 / 66	6 / 22	43	64	38	
United Arab Emirates	14	73.3 / 77.4	54	94 / 90	97 / 98	77 / 82	24 / 19	51	28	24	
Yemen	71	58.9 / 61.1	570	97 / 64	82 / 94	65 / 27	31 / 71	111	21	10	
ARAB STATES (11)	**53**	**63.9 / 67.1**	**252**	**96 / 88**	**93 / 94**	**68 / 62**	**26 / 48**	**48**	**40**	**34**	**0.43 / 0.55**
EUROPE	**9**	**70.1 / 78.2**						**20**	**67**	**49**	
EASTERN EUROPE	14	64.0 / 74.4						29	61	36	
Bulgaria	15	67.4 / 74.6	32	103 / 100	95 / 95	94 / 91	1 / 2	41	42	25	
Czech Republic	6	72.1 / 78.7	9	104 / 103	96 / 97	95 / 97		17	72	63	0.1 / <0.1
Hungary	9	67.7 / 76.0	16	102 / 100	98 / 99	103 / 104		21	77	68	
Poland	9	69.8 / 78.0	13	100 / 99	99 / 98	105 / 101		16	49	19	
Romania	20	67.0 / 74.2	49	100 / 98	94 / 95	82 / 83	2 / 4	37	64	30	
Slovakia	8	69.8 / 77.6	3	102 / 101	98 / 99	89 / 90	0 / 0	24	74	41	
NORTHERN EUROPE (12)	5	74.9 / 80.5						17	79	75	
Denmark	5	74.2 / 79.1	5	102 / 102	100 / 100			7	78	72	0.3 / 0.1
Estonia	9	66.5 / 76.8	63	105 / 101	100 / 99	109 / 111	0 / 0	26	70	56	1.4 / 0.7
Finland	4	74.4 / 81.5	6	102 / 101	100 / 100	120 / 133		8	77	75	0.1 / <0.1
Ireland	6	74.4 / 79.6	5	104 / 104	98 / 99	100 / 109		15			0.2 / 0.1
Latvia	14	65.6 / 76.2	42	99 / 98	98 / 98	92 / 93	0 / 0	24	48	39	0.8 / 0.4
Lithuania	9	67.5 / 77.6	13	105 / 104	100 / 98	99 / 98	0 / 0	26	47	31	0.1 / 0.0
Norway	5	76.0 / 81.9	16	101 / 102		113 / 116		11	74	69	0.1 / 0.0
Sweden	3	77.6 / 82.6	2	109 / 112		132 / 160		7	78	72	0.1 / 0.0
United Kingdom	5	75.7 / 80.7	13	101 / 101		146 / 170		20	84	81	0.2 / 0.0
SOUTHERN EUROPE (13)	7	74.6 / 81.0						11	67	46	
Albania	25	70.9 / 76.7	55	107 / 107	86 / 94	77 / 80	1 / 2	16	58	15	
Bosnia and Herzegovina	14	71.3 / 76.7	31				2 / 9	23	48	16	
Croatia	8	70.3 / 78.1	8	96 / 95	100 / 100	88 / 89	1 / 3	19			
Greece	6	75.7 / 80.9	9	97 / 96		95 / 97		10			0.3 / 0.1
Italy	5	75.5 / 81.9	5	101 / 100	96 / 97	97 / 95		6	60	39	0.7 / 0.3
Macedonia (Former Yugoslav Republic of)	16	71.4 / 75.8	23	99 / 99	96 / 97	86 / 83		34			
Portugal	6	72.6 / 79.6	5	122 / 120		111 / 117		17	66	33	0.7 / 0.2
Serbia and Montenegro	13	70.9 / 75.6	11	99 / 99		88 / 89		26	58	33	0.3 / 0.1
Slovenia	6	72.6 / 79.8	17	101 / 100		105 / 107	0 / 0	8	74	59	
Spain	5	75.9 / 82.8	4	108 / 106		112 / 119		6	81	67	1.0 / 0.3
WESTERN EUROPE (14)	5	75.3 / 81.7						10	74	71	
Austria	5	75.4 / 81.5	4	104 / 103	93 / 95	100 / 97		12	51	47	0.4 / 0.1
Belgium	4	75.7 / 81.9	10	106 / 105		146 / 163		9	78	74	0.3 / 0.1
France	5	75.2 / 82.8	17	106 / 104		107 / 108		9	75	69	0.6 / 0.2
Germany	5	75.2 / 81.2	8	101 / 100	99 / 100	100 / 99		11	75	72	0.2 / 0.0
Netherlands	5	75.6 / 81.0	16	109 / 107	100 / 100	126 / 122		5	79	76	0.4 / 0.1
Switzerland	5	75.9 / 82.3	7	108 / 107	100 / 99	103 / 96		5	82	78	0.5 / 0.2
LATIN AMERICA & CARIBBEAN	**32**	**67.1 / 73.9**						**72**	**71**	**62**	
CARIBBEAN (15)	35	64.9 / 69.0						71	61	57	
Cuba	7	74.8 / 78.7	33	102 / 98	95 / 96	90 / 89	3 / 3	65	73	72	0.1 / <0.1
Dominican Republic	36	64.4 / 69.2	150	125 / 127	54 / 80	60 / 75	16 / 16	93	65	63	2.5 / 0.9

	Indicators of Mortality			Indicators of Education				Reproductive Health Indicators			
	Infant mortality Total per 1,000 live births	Life expectancy M/F	Maternal mortality ratio	Primary enrolment (gross) M/F	Proportion reaching grade 5 M/F	Secondary enrolment (gross) M/F	% Illiterate (>15 years) M/F	Births per 1,000 women aged 15-19	Contraceptive Prevalence Any method	Contraceptive Prevalence Modern methods	HIV prevalence rate (%) (15-49) M/F
Haiti	63	49.0 / 50.0	680				46 / 50	64	27	21	4.8 / 6.4
Jamaica	20	73.7 / 77.8	87	101 / 100	88 / 93	82 / 85	16 / 9	79	66	63	1.2 / 1.2
Puerto Rico	10	71.2 / 80.1	25				6 / 6	63	78	68	
Trinidad and Tobago	14	68.4 / 74.4	160	106 / 104	97 / 100	69 / 73	1 / 2	36	38	33	3.2 / 3.2
CENTRAL AMERICA	30	69.5 / 75.4						76	64	55	
Costa Rica	10	75.8 / 80.6	43	108 / 108	93 / 95	66 / 68	4 / 4	78	75	65	0.8 / 0.4
El Salvador	26	67.7 / 73.7	150	114 / 109	65 / 70	56 / 56	18 / 23	87	60	54	0.9 / 0.5
Guatemala	41	63.0 / 68.9	240	107 / 99	57 / 54	34 / 32	23 / 38	111	38	31	1.3 / 1.0
Honduras	32	66.5 / 71.4	110	105 / 107			20 / 20	103	62	51	1.7 / 2.0
Mexico	28	70.4 / 76.4	83	111 / 110	90 / 91	73 / 78	7 / 11	64	67	58	0.3 / 0.2
Nicaragua	36	67.2 / 71.9	230	104 / 105	51 / 58	52 / 61	23 / 23	135	69	66	0.3 / 0.1
Panama	21	72.3 / 77.4	160	112 / 108	88 / 89	67 / 72	7 / 8	89	58	54	1.1 / 0.8
SOUTH AMERICA (16)	32	66.5 / 73.9						71	74	65	
Argentina	20	70.6 / 77.7	82	120 / 119	91 / 95	97 / 103	3 / 3	61			1.1 / 0.3
Bolivia	56	61.8 / 66.0	420	114 / 113	79 / 77	86 / 83	7 / 19	81	53	27	0.1 / 0.1
Brazil	38	64.0 / 72.6	260	153 / 144	76 / 84	102 / 113	14 / 13	73	77	70	0.8 / 0.5
Chile	12	73.0 / 79.0	31	101 / 99	100 / 100	88 / 90	4 / 4	44			0.4 / 0.2
Colombia	26	69.2 / 75.3	130	110 / 109	59 / 63	62 / 69	8 / 8	80	77	64	1.0 / 0.5
Ecuador	41	68.3 / 73.5	130	117 / 117	77 / 79	59 / 59	8 / 10	66	66	50	0.4 / 0.2
Paraguay	37	68.6 / 73.1	170	114 / 110	76 / 78	63 / 64	7 / 10	75	57	48	0.7 / 0.3
Peru	33	67.3 / 72.4	410	120 / 120	86 / 86	92 / 86	9 / 20	55	69	50	0.7 / 0.4
Uruguay	13	71.6 / 78.9	27	109 / 107	87 / 90	95 / 108	3 / 2	70			0.4 / 0.2
Venezuela	19	70.9 / 76.7	96	107 / 105	92 / 100	64 / 74	6 / 7	95			0.9 / 0.5
NORTHERN AMERICA (17)	7	74.5 / 80.1						50	76	71	
Canada	5	76.7 / 81.9	6	99 / 100		107 / 106		16	75	73	0.5 / 0.2
United States of America	7	74.3 / 79.9	17	98 / 99		94 / 92		53	76	71	1.0 / 0.3
OCEANIA	26	71.8 / 76.6						32	62	57	
AUSTRALIA-NEW ZEALAND	6	76.3 / 81.8						17	76	72	
Australia (18)	6	76.4 / 82.0	8	102 / 102		155 / 153		16	76	72	0.2 / <0.1
Melanesia (19)	53	59.3 / 61.7						63			
New Zealand	6	75.8 / 80.7	7	99 / 99		109 / 118		27	75	72	0.1 / <0.1
Papua New Guinea	62	56.8 / 58.7	300	77 / 78	61 / 58	25 / 20		67	26	20	0.8 / 0.4
COUNTRIES WITH ECONOMIES IN TRANSITION OF THE FORMER USSR (20)											
Armenia	17	69.0 / 75.6	55	97 / 95		84 / 89	0 / 1	34	61	22	0.1 / 0.1
Azerbaijan	29	68.7 / 75.5	94	93 / 92	96 / 99	81 / 79		36	55	12	
Belarus	11	64.9 / 75.3	35	111 / 109		82 / 86	0 / 0	27	50	42	
Georgia	18	69.5 / 77.6	32	92 / 92	94 / 94	76 / 82		33	41	20	0.3 / 0.1
Kazakhstan	52	60.9 / 71.9	210	100 / 99	95 / 95	90 / 88	0 / 1	45	66	53	0.2 / 0.1
Kyrgyzstan	37	64.8 / 72.3	110	102 / 99	91 / 91	86 / 87		33	60	49	0.1 / <0.1
Republic of Moldova	18	65.5 / 72.2	36	86 / 85	90 / 91	71 / 73	0 / 1	43	62	43	
Russian Federation	16	60.8 / 73.1	67	114 / 113		92 / 92	0 / 1	30			1.5 / 0.8
Tajikistan	50	66.2 / 71.4	100	109 / 104	93 / 100	90 / 74	0 / 1	25	34	27	
Turkmenistan	49	63.9 / 70.4	31				1 / 2	17	62	53	
Ukraine	14	64.7 / 74.7	35	91 / 90	30 / 30	97 / 97	0 / 0	38	68	38	1.8 / 0.9
Uzbekistan	37	66.8 / 72.5	24	103 / 102		100 / 97	0 / 1	54	67	63	0.1 / 0.1

Demographic, Social and Economic Indicators

	Total population (millions) (2004)	Projected population (millions) (2050)	Ave. pop. growth rate (%) (2000-2005)	% urban (2003)	Urban growth rate (2000-2005)	Population/ ha arable & perm. crop land	Total fertility rate (2000-2005)	% births with skilled atten-dants	GNI per capita PPP$ (2002)	Expen-ditures/ primary student (% of GDP per capita)	Health expen-ditures, public (% of GDP)	External population assistance (US$,000)	Under-5 mortality M/F	Per capita energy con-sumption	Access to safe water
World Total	**6,377.6**	**8,918.7**	**1.2**	**48**	**2.1**		**2.69**					**(2,521,000)**	**81 / 81**		
More developed regions (*)	**1,206.1**	**1,219.7**	**0.2**	**75**	**0.5**		**1.56**						**10 / 9**		
Less developed regions (+)	**5,171.5**	**7,699.1**	**1.5**	**42**	**2.8**		**2.92**						**89 / 89**		
Least developed countries (‡)	**735.6**	**1,674.5**	**2.4**	**27**	**4.3**		**5.13**						**165 / 156**		
AFRICA (1)	**869.2**	**1,803.3**	**2.2**	**39**	**3.6**		**4.91**					**605,466**[21]	**154 / 143**		
EASTERN AFRICA	**276.2**	**614.5**	**2.2**	**26**	**4.3**		**5.61**						**171 / 156**		
Burundi	7.1	19.5	3.1	10	6.5	4.6	6.80	25	610	11.6	2.1	2,255	198 / 178		78
Eritrea	4.3	10.5	3.7	20	5.8	5.9	5.43	28	950		3.7	6,774	108 / 104		46
Ethiopia	72.4	171.0	2.5	16	4.1	4.8	6.14	6	720		1.4	43,125	181 / 165	291	24
Kenya	32.4	44.0	1.5	39	4.4	4.5	4.00	41	990	0.9	1.7	38,134	125 / 110	500	57
Madagascar	17.9	46.3	2.8	27	3.6	3.4	5.70	46	720	10.7	1.2	10,208	150 / 144		47
Malawi	12.3	25.9	2.0	16	4.6	3.8	6.10	56	570		2.7	22,230	192 / 181		57
Mauritius (2)	1.2	1.5	1.0	43	1.5	1.3	1.95	99	10,530	9.0	2.0	193	21 / 15		100
Mozambique	19.2	31.3	1.8	36	5.1	3.3	5.63	44			4.0	29,800	223 / 207	425	57
Rwanda	8.5	17.0	2.2	18	11.6	5.6	5.74	31	1,210	6.9	3.1	14,044	189 / 168		41
Somalia	10.3	39.7	4.2	35	5.7	6.0	7.25	34			1.2		203 / 187		
Uganda	26.7	103.2	3.2	12	3.9	2.6	7.10	39	1,320		3.4	42,399	154 / 139		52
United Republic of Tanzania	37.7	69.1	1.9	35	4.9	5.6	5.11	36	550		2.0	31,019	170 / 153	404	68
Zambia	10.9	18.5	1.2	36	1.9	1.4	5.64	43	770		3.0	29,312	194 / 177	638	64
Zimbabwe	12.9	12.7	0.5	35	1.8	2.4	3.90	73	2,120	16.2	2.8	17,364	118 / 109	769	83
MIDDLE AFRICA (3)	**103.4**	**266.3**	**2.7**	**37**	**4.1**		**6.28**						**218 / 196**		
Angola	14.1	43.1	3.2	36	5.4	2.8	7.20	45	1,730		2.8	8,057	259 / 234	663	38
Cameroon	16.3	24.9	1.8	51	3.4	1.1	4.61	60	1,640	8.5	1.2	3,343	155 / 142	417	58
Central African Republic	3.9	6.6	1.3	43	2.5	1.3	4.92	44	1,190		2.3	982	189 / 157		70
Chad	8.9	25.4	3.0	25	4.6	1.7	6.65	16	1,000	9.5	2.0	2,675	209 / 192		27
Congo, Democratic Republic of the (4)	54.4	151.6	2.9	32	4.4	4.0	6.70	61	580		1.5	8,783	230 / 208	300	45
Congo, Republic of	3.8	10.6	2.6	54	3.4	6.4	6.29		700	0.4	1.4	928	137 / 113	262	51
Gabon	1.4	2.5	1.8	84	2.7	0.9	3.99	86	5,320	4.7	1.7	3,069	97 / 87	1,322	86
NORTHERN AFRICA (5)	**187.0**	**306.0**	**1.9**	**50**	**2.7**		**3.21**					**79,135**[22]	**70 / 61**		
Algeria	32.3	48.7	1.7	59	2.6	0.9	2.80	92	5,330		3.1	3,492	52 / 45	955	89
Egypt	73.4	127.4	2.0	42	2.1	7.5	3.29	61	3,710		1.9	58,689	52 / 44	737	97
Libyan Arab Jamahiriya	5.7	9.2	1.9	86	2.3	0.1	3.02	94			1.6	0	23 / 23	2,994	72
Morocco	31.1	47.1	1.6	58	2.8	1.1	2.75	40	3,690	17.9	2.0	9,699	58 / 46	377	80
Sudan	34.3	60.1	2.2	39	4.6	1.2	4.39		1,690		0.6	5,261	131 / 123	421	75
Tunisia	9.9	12.9	1.1	64	1.6	0.5	2.01	90	6,280	15.8	4.9	1,069	29 / 24	852	80
SOUTHERN AFRICA	**51.9**	**46.6**	**0.6**	**54**	**1.5**		**2.79**						**93 / 83**		
Botswana	1.8	1.4	0.9	52	1.8	2.1	3.70	94	7,770	6.0	4.4	2,692	108 / 100		95
Lesotho	1.8	1.4	0.1	18	0.9	2.1	3.84	60	2,710	21.4	4.3	967	158 / 146		78
Namibia	2.0	2.7	1.4	32	3.0	1.1	4.56	78	6,650	22.1	4.7	4,080	113 / 102	596	77
South Africa	45.2	40.2	0.6	57	1.4	0.4	2.61	84	9,870	14.3	3.6	29,267	85 / 75	2,404	86
Swaziland	1.1	0.9	0.8	24	1.4	1.9	4.54	70	4,530	10.4	2.3	635	155 / 138		
WESTERN AFRICA (6)	**250.6**	**569.9**	**2.6**	**42**	**4.2**		**5.56**						**153 / 148**		
Benin	6.9	15.6	2.6	45	4.4	1.5	5.66	66	1,020	10.1	2.1	7,766	166 / 146	318	63
Burkina Faso	13.4	42.4	3.0	18	5.0	2.8	6.68	31	1,010		1.5	6,691	165 / 155		42
Côte d'Ivoire	16.9	27.6	1.6	45	2.6	1.0	4.73	63	1,430	14.9	1.0	4,014	182 / 164	402	81
Gambia	1.5	2.9	2.7	26	2.6	4.2	4.70	55	1,680		3.2	690	140 / 128		62

Demographic, Social and Economic Indicators

	Total population (millions) (2004)	Projected population (millions) (2050)	Ave. pop. growth rate (%) (2000-2005)	% urban (2003)	Urban growth rate (2000-2005)	Population/ ha arable & perm. crop land	Total fertility rate (2000-2005)	% births with skilled attendants	GNI per capita PPP$ (2002)	Expenditures/ primary student (% of GDP per capita)	Health expenditures, public (% of GDP)	External population assistance (US$,000)	Under-5 mortality M/F	Per capita energy consumption	Access to safe water
Ghana	21.4	39.5	2.2	45	3.2	1.9	4.11	44	2,000		2.8	21,753	99 / 88	410	73
Guinea	8.6	19.6	1.6	35	3.8	4.5	5.82	35	1,990	9.2	1.9	6,176	175 / 176		48
Guinea-Bissau	1.5	4.7	2.9	34	5.4	2.1	7.10	35	750		3.2	562	221 / 198		56
Liberia	3.5	9.8	4.0	47	5.3	3.5	6.80	51			3.3	1,626	238 / 221		
Mali	13.4	46.0	3.0	32	5.2	2.1	7.00	41	840	14.4	1.7	14,171	184 / 178		65
Mauritania	3.0	7.5	3.0	62	5.1	2.9	5.79	57	1,740	14.0	2.6	2,061	163 / 150		37
Niger	12.4	53.0	3.6	22	6.1	2.2	8.00	16	770	16.8	1.4	3,979	207 / 213		59
Nigeria	127.1	258.5	2.5	47	4.4	1.2	5.42	35	780		0.8	35,933	133 / 133	735	62
Senegal	10.3	21.6	2.4	50	3.9	2.8	4.97	58	1,510	13.8	2.8	17,082	116 / 108	325	78
Sierra Leone	5.2	10.3	3.8	39	5.6	5.0	6.50	42	490		2.6	889	321 / 293		57
Togo	5.0	10.0	2.3	35	4.0	1.1	5.33	49	1,430	11.0	1.5	2,695	145 / 128	305	54
ASIA	**3,870.5**	**5,222.1**	**1.3**	**39**	**2.7**		**2.55**					**396,994**	**68 / 73**		
EASTERN ASIA (7)	1,522.0	1,590.1	0.7	43	2.6		1.78						36 / 44		
China	1,313.3	1,395.2	0.7	39	3.2	5.5	1.83	76	4,390	6.6	2.0	22,176	39 / 47	896	75
Democratic People's Republic of Korea	22.8	25.0	0.5	61	1.0	2.4	2.02	97			1.9	1,198	61 / 55	914	100
Hong Kong SAR, China (8)	7.1	9.4	1.1	100	1.1		1.00		26,810				5 / 5	2,421	
Japan	127.8	109.7	0.1	65	0.3	1.0	1.32	100	26,070	21.4	6.2	(115,346) [23]	5 / 4	4,099	
Mongolia	2.6	3.8	1.3	57	1.4	0.5	2.42	97	1,650		4.6	3,989	88 / 83		60
Republic of Korea	48.0	46.4	0.6	80	0.9	2.1	1.41	100	16,480	18.4	2.6	0	8 / 6	4,114	92
SOUTH-EASTERN ASIA	550.7	767.2	1.4	42	3.3		2.55						61 / 49		
Cambodia	14.5	29.6	2.4	19	5.5	2.5	4.77	32	1,590	7.4	1.7	24,787	115 / 99		30
Indonesia	222.6	293.8	1.3	46	3.9	2.8	2.35	66	2,990	3.7	0.6	34,244	59 / 46	729	78
Lao People's Democratic Republic	5.8	11.4	2.3	21	4.6	4.3	4.78	19	1,610	9.1	1.7	2,244	144 / 137		37
Malaysia	24.9	39.6	1.9	64	3.0	0.5	2.90	97	8,280	17.0	2.0	156	15 / 11	2,168	
Myanmar	50.1	64.5	1.3	29	3.1	3.2	2.86	56		5.8	0.4	4,688	137 / 118	252	72
Philippines	81.4	127.0	1.8	61	3.1	2.8	3.18	58	4,280	11.8	1.5	46,523	40 / 30	538	86
Singapore	4.3	4.5	1.7	100	1.7	5.9	1.36	100	23,090		1.3	0	4 / 4	7,058	100
Thailand	63.5	77.1	1.0	32	1.9	1.6	1.93	99	6,680	15.9	2.1	2,466	31 / 19	1,235	84
Viet Nam	82.5	117.7	1.3	26	3.2	6.3	2.30	85	2,240		1.5	16,392	52 / 37	495	77
SOUTH CENTRAL ASIA	1,588.8	2,463.9	1.7	30	2.5		3.25						89 / 98		
Afghanistan	24.9	69.5	3.9	23	6.0	1.8	6.80	12			2.7	1,491	278 / 283		13
Bangladesh	149.7	254.6	2.0	24	3.5	9.1	3.46	12	1,720	8.3	1.5	75,909	85 / 90	153	97
Bhutan	2.3	5.3	3.0	9	6.3	12.1	5.02	24			3.6	632	82 / 78		62
India	1,081.2	1,531.4	1.5	28	2.3	3.2	3.01	43	2,570	13.7	0.9	57,199	78 / 90	515	84
Iran (Islamic Republic of)	69.8	105.5	1.2	67	2.3	1.1	2.33	90	6,340	11.6	2.7	2,276	39 / 39	1,860	92
Nepal	25.7	50.8	2.2	15	5.2	7.0	4.26	11	1,350	12.5	1.5	19,820	91 / 106	357	88
Pakistan	157.3	348.7	2.4	34	3.4	3.3	5.08	20	1,940		1.0	13,415	121 / 135	456	90
Sri Lanka	19.2	21.2	0.8	21	0.7	4.5	2.01	97	3,390	10.0	1.8	2,074	30 / 16	423	77
WESTERN ASIA	208.9	400.8	2.1	65	2.4		3.45					30,221 [22]	60 / 53		
Iraq	25.9	57.9	2.7	67	2.4	0.4	4.77	72			1.0	268	112 / 103	1,202	85
Israel	6.6	10.0	2.0	92	2.1	0.4	2.70			21.0	6.0	0	9 / 9	3,291	
Jordan	5.6	10.2	2.7	79	2.8	1.4	3.57	100	4,070	16.0	4.5	14,233	28 / 26	1,017	96
Kuwait	2.6	4.9	3.5	96	3.5	1.7	2.66	98			3.5	0	13 / 13	7,195	
Lebanon	3.7	4.9	1.6	88	1.9	0.4	2.18	89	4,470	8.3	2.2	1,885	22 / 17	1,239	100
Occupied Palestinian Territory	3.7	11.1	3.6	71	4.1		5.57	97				2,385	27 / 21		86
Oman	2.9	6.8	2.9	78	3.6	12.0	4.96	95	12,910	12.6	2.4	77	26 / 20	4,029	39

Demographic, Social and Economic Indicators

	Total population (millions) (2004)	Projected population (millions) (2050)	Ave. pop. growth rate (%) (2000-2005)	% urban (2003)	Urban growth rate (2000-2005)	Population/ ha arable & perm. crop land	Total fertility rate (2000-2005)	% births with skilled atten-dants	GNI per capita PPP$ (2002)	Expen-ditures/ primary student (% of GDP per capita)	Health expen-ditures, public (% of GDP)	External population assistance (US$,000)	Under-5 mortality M/F	Per capita energy con-sumption	Access to safe water
Saudi Arabia	24.9	54.7	2.9	88	3.4	0.6	4.53	91		34.9	3.4	4	26 / 23	5,195	95
Syrian Arab Republic	18.2	34.2	2.4	50	2.5	0.9	3.32		3,250	12.8	1.7	3,063	28 / 25	841	80
Turkey (10)	72.3	97.8	1.4	66	2.2	0.8	2.43	81	6,120	11.6	4.4	2,650	56 / 43	1,057	82
United Arab Emirates	3.1	4.1	1.9	85	2.1	0.6	2.82	96		9.2	2.6	0	17 / 14	10,860	
Yemen	20.7	84.4	3.5	26	4.8	5.8	7.01	22	750		1.5	5,647	100 / 95	197	69
ARAB STATES (11)	**314.2**	**631.2**	**2.3**	**55**	**2.9**	**1.1**	**3.81**	**67**	**3,547**	**23.9**	**2.7**	**109,243**	**77 / 70**	**1,400**	**85**
EUROPE	**725.6**	**631.9**	**-0.1**	**73**	**0.1**		**1.38**						**12 / 10**		
EASTERN EUROPE	**298.8**	**221.7**	**-0.5**	**68**	**-0.4**		**1.18**					**35,259** [22, 24]	**20 / 16**		
Bulgaria	7.8	5.3	-0.8	70	-0.3	0.1	1.10		6,840	14.8	3.9	155	21 / 17	2,428	100
Czech Republic	10.2	8.6	-0.1	74	0.0	0.2	1.16	99	14,500	13.0	6.7	197	6 / 6	4,049	
Hungary	9.8	7.6	-0.5	65	0.1	0.2	1.20		12,810	19.2	5.1	0	12 / 10	2,487	99
Poland	38.6	33.0	-0.1	62	0.0	0.5	1.26		10,130	28.8	4.6	109	11 / 10	2,344	
Romania	22.3	18.1	-0.2	55	-0.2	0.3	1.32	98	6,290		5.2	4,414	28 / 22	1,644	58
Slovakia	5.4	4.9	0.1	57	0.5	0.3	1.28		12,190	11.4	5.1	17	10 / 10	3,480	100
NORTHERN EUROPE (12)	**95.0**	**100.1**	**0.2**	**83**	**0.4**		**1.61**						**7 / 6**		
Denmark	5.4	5.3	0.2	85	0.3	0.1	1.77		29,450	23.4	7.0	(48,852)	7 / 6	3,692	100
Estonia	1.3	0.7	-1.1	69	-1.0	0.2	1.22		11,120	23.6	4.3	50	13 / 9	3,444	
Finland	5.2	4.9	0.2	61	0.1	0.1	1.73		25,440		5.3	(23,730)	5 / 4	6,518	100
Ireland	4.0	5.0	1.1	60	1.5	0.4	1.90	100	28,040		4.9	(6,255)	7 / 7	3,876	
Latvia	2.3	1.3	-0.9	66	-1.2	0.1	1.10	100	8,940	23.1	3.4	93	19 / 16	1,822	
Lithuania	3.4	2.5	-0.6	67	-0.7	0.2	1.25		9,880		4.2	85	13 / 10	2,304	
Norway	4.6	4.9	0.4	79	1.6	0.3	1.80		35,840	26.8	6.8	(42,960)	6 / 5	5,896	100
Sweden	8.9	8.7	0.1	83	0.1	0.1	1.64		25,080	24.3	7.4	(56,270)	5 / 4	5,740	100
United Kingdom	59.4	66.2	0.3	89	0.4	0.2	1.60	99	25,870	13.6	6.3	(80,971)	7 / 6	3,982	100
SOUTHERN EUROPE (13)	**146.4**	**125.6**	**0.1**	**66**	**0.3**		**1.32**						**10 / 9**		
Albania	3.2	3.7	0.7	44	2.1	2.1	2.28	99	4,040		2.4	1,928	37 / 31	548	97
Bosnia and Herzegovina	4.2	3.6	1.1	44	2.2	0.2	1.30	100	5,800		2.8	175	17 / 14	1,074	
Croatia	4.4	3.6	-0.2	59	0.5	0.2	1.65	100	9,760		7.3	0	10 / 8	1,771	
Greece	11.0	9.8	0.1	61	0.6	0.4	1.27		18,240		5.2		8 / 7	2,710	
Italy	57.3	44.9	-0.1	67	0.0	0.3	1.23		25,320		6.3	(25,038)	7 / 6	2,981	
Macedonia (Former Yugoslav Republic of)	2.1	2.2	0.5	60	0.6	0.4	1.90	97	6,210	16.6	5.8		19 / 18		
Portugal	10.1	9.0	0.1	55	1.1	0.5	1.45	100	17,350		6.3	(689)	9 / 8	2,435	
Serbia and Montenegro	10.5	9.4	-0.1	52	0.2	0.5	1.65	99			6.5	1,780	17 / 14	1,508	98
Slovenia	2.0	1.6	-0.1	51	-0.1	0.2	1.14		17,690		6.3	0	8 / 7	3,459	100
Spain	41.1	37.3	0.2	77	0.3	0.2	1.15		20,460		5.4	(14,380)	7 / 6	3,127	
WESTERN EUROPE (14)	**185.3**	**184.5**	**0.2**	**81**	**0.5**		**1.58**						**6 / 6**		
Austria	8.1	7.4	0.0	66	0.0	0.3	1.28		28,240		5.5	(979)	6 / 5	3,825	100
Belgium	10.3	10.2	0.0	97	0.2		1.66		27,350		6.4	(19,066)	6 / 6	5,735	
France	60.4	64.2	0.5	76	0.7	0.1	1.89		26,180		7.3	(8,242)	6 / 6	4,487	
Germany	82.5	79.1	0.1	88	0.3	0.2	1.35		26,220		8.1	(108,660) [25]	6 / 6	4,264	
Netherlands	16.2	17.0	0.5	66	1.3	0.6	1.72	100	27,470		5.7	(132,032)	7 / 6	4,814	100
Switzerland	7.2	5.8	0.0	68	-0.1	1.0	1.41		31,250	22.8	6.4	(23,534)	7 / 5	3,875	100
LATIN AMERICA & CARIBBEAN	**550.8**	**767.7**	**1.4**	**77**	**1.9**		**2.53**					**188,603**	**45 / 36**		
CARIBBEAN (15)	**39.0**	**45.8**	**0.9**	**64**	**1.3**		**2.39**						**62 / 53**		
Cuba	11.3	10.1	0.3	76	0.5	0.4	1.55	100		32.7	6.2	1,469	12 / 8	1,216	91
Dominican Republic	8.9	11.9	1.5	59	2.1	0.9	2.71	98	5,870	6.6	2.2	8,135	58 / 48	921	86

	Total population (millions) (2004)	Projected population (millions) (2050)	Ave. pop. growth rate (%) (2000-2005)	% urban (2003)	Urban growth rate (2000-2005)	Population/ha arable & perm. crop land	Total fertility rate (2000-2005)	% births with skilled attendants	GNI per capita PPP$ (2002)	Expenditures/primary student (% of GDP per capita)	Health expenditures, public (% of GDP)	External population assistance (US$,000)	Under-5 mortality M/F	Per capita energy consumption	Access to safe water
Haiti	8.4	12.4	1.3	38	3.0	4.6	3.98	24	1,580		2.7	16,621	119 / 104	257	46
Jamaica	2.7	3.7	0.9	52	1.0	1.9	2.36	95	3,550	15.7	2.9	3,534	28 / 21	1,545	92
Puerto Rico	3.9	3.7	0.5	97	1.1	1.3	1.89					0	14 / 11		
Trinidad and Tobago	1.3	1.2	0.3	75	0.9	0.9	1.55	96	8,680	14.2	1.7	530	21 / 16	6,708	90
CENTRAL AMERICA	144.6	211.8	1.7	69	2.0		2.76						41 / 34		
Costa Rica	4.3	6.5	1.9	61	2.8	1.6	2.28	98	8,260	14.6	4.9	344	14 / 11	899	95
El Salvador	6.6	9.8	1.6	60	2.1	2.2	2.88	90	4,570		3.7	7,760	38 / 31	677	77
Guatemala	12.7	26.2	2.6	46	3.4	3.0	4.41	41	3,880	7.7	2.3	12,474	58 / 51	626	92
Honduras	7.1	12.6	2.3	46	3.3	1.6	3.72	56	2,450		3.2	13,853	53 / 43	488	88
Mexico	104.9	140.2	1.5	76	1.8	0.8	2.50	86	8,540	11.8	2.7	9,849	37 / 31	1,532	88
Nicaragua	5.6	10.9	2.4	57	3.1	0.5	3.75	67		20.5	3.8	16,685	50 / 40	536	77
Panama	3.2	5.1	1.8	57	2.4	1.0	2.70	90	5,870	10.5	4.8	474	31 / 23	1,098	90
SOUTH AMERICA (16)	367.2	510.1	1.4	81	2.0		2.45						45 / 35		
Argentina	38.9	52.8	1.2	90	1.4	0.1	2.44	98	9,930	12.4	5.1	865	26 / 21	1,537	
Bolivia	9.0	15.7	1.9	63	2.7	1.2	3.82	69	2,300	12.0	3.5	25,576	77 / 67	496	83
Brazil	180.7	233.1	1.2	83	2.0	0.4	2.21	88	7,250	10.7	3.2	7,545	52 / 39	1,074	87
Chile	16.0	21.8	1.2	87	1.6	1.0	2.35	100	9,180	14.3	3.1	112	15 / 12	1,545	93
Colombia	44.9	67.5	1.6	77	2.2	2.0	2.62	86	5,870	16.4	3.6	1,427	35 / 30	680	91
Ecuador	13.2	18.7	1.5	62	2.3	1.1	2.76	69	3,222		2.3	9,697	60 / 49	692	85
Paraguay	6.0	12.1	2.4	57	3.5	0.7	3.84	71	4,450	12.9	3.0	3,761	51 / 39	697	78
Peru	27.6	41.1	1.5	74	2.0	1.9	2.86	59	4,800	7.5	2.6	23,635	57 / 47	460	80
Uruguay	3.4	4.1	0.7	93	0.9	0.3	2.30	100	12,010	7.2	5.1	193	18 / 13	809	98
Venezuela	26.2	41.7	1.9	88	2.1	0.7	2.72	94	5,080		3.7	879	25 / 20	2,227	83
NORTHERN AMERICA (17)	328.9	447.9	1.0	80	1.4		2.05						8 / 8		
Canada	31.7	39.1	0.8	80	1.2	0.0	1.48	98	28,070		6.8	(12,689)	7 / 6	7,985	100
United States of America	297.0	408.7	1.0	80	1.4	0.0	2.11	99	35,060	18.0	6.2	(951,012)	8 / 9	7,996	100
OCEANIA	32.6	45.8	1.2	73	1.4		2.34						34 / 35		
AUSTRALIA-NEW ZEALAND	23.8	30.1	0.9	91	1.3		1.75						8 / 6		
Australia (18)	19.9	25.6	1.0	92	1.4	0.0	1.70	100	26,960	16.0	6.2	(13,088)	8 / 6	5,956	100
Melanesia (19)	7.6	14.0	2.1	20	2.5		3.91						70 / 75		
New Zealand	3.9	4.5	0.8	86	0.8	0.1	2.01	100	20,020	19.6	6.4	(2,150)	8 / 6	4,714	
Papua New Guinea	5.8	11.1	2.2	13	2.3	4.9	4.09	53	2,080	12.4	3.9	6,157	81 / 88		42
COUNTRIES WITH ECONOMIES IN TRANSITION OF THE FORMER USSR (20)															
Armenia	3.1	2.3	-0.5	64	-0.8	0.7	1.15	97	3,060		3.2	3,721	22 / 17	744	
Azerbaijan	8.4	10.9	0.9	50	0.6	1.1	2.10	84	2,920		0.7	1,887	41 / 38	1,428	78
Belarus	9.9	7.5	-0.5	71	0.1	0.2	1.20	100	5,330		4.8	148	17 / 12	2,449	100
Georgia	5.1	3.5	-0.9	52	-1.4	1.0	1.40	96	2,210		1.4	2,991	25 / 18	462	79
Kazakhstan	15.4	13.9	-0.4	56	-0.3	0.1	1.95	99	5,480		1.9	6,169	68 / 48	2,705	91
Kyrgyzstan	5.2	7.2	1.4	34	1.0	0.9	2.64	98	1,520		1.9	2,593	50 / 42	451	77
Republic of Moldova	4.3	3.6	-0.1	46	0.1	0.4	1.40	99	1,560		2.8	768	26 / 21	735	92
Russian Federation	142.4	101.5	-0.6	73	-0.6	0.1	1.14	99	7,820		3.7	12,226	23 / 18	4,293	99
Tajikistan	6.3	9.6	0.9	25	-0.4	1.9	3.06	71	900		1.0	805	78 / 67	487	60
Turkmenistan	4.9	7.5	1.5	45	2.0	0.9	2.70	97	4,570		3.0	1,027	74 / 61	3,244	
Ukraine	48.2	31.7	-0.8	67	-0.7	0.2	1.15	100	4,650		2.9	4,658	20 / 15	2,884	98
Uzbekistan	26.5	37.8	1.5	37	1.0	1.4	2.44	96	1,590		2.7	4,624	56 / 48	2,029	85

Selected Indicators for Less Populous Countries/Territories

Monitoring ICPD Goals – Selected Indicators	Indicators of Mortality			Indicators of Education		Reproductive Health Indicators			
	Infant mortality Total per 1,000 live births	Life expectancy M/F	Maternal mortality ratio	Primary enrolment (gross) M/F	Secondary enrolment (gross) M/F	Births per 1,000 women aged 15-19	Contraceptive Prevalence Any method	Modern methods	HIV prevalence rate (%) (15-49) M/F
Bahamas	18	63.9 / 70.3	60	92 / 93	90 / 93	60	62	60	3.0 / 3.0
Bahrain	14	72.5 / 75.9	28	98 / 98	91 / 99	18	62	31	0.1 / 0.2
Barbados	11	74.5 / 79.5	95	108 / 108	103 / 103	43	55	53	2.0 / 1.0
Belize	31	69.9 / 73.0	140	119 / 116	68 / 74	86	47	42	3.0 / 1.8
Brunei Darussalam	6	74.2 / 78.9	37	107 / 106	85 / 91	26			<0.1 / <0.1
Cape Verde	30	67.0 / 72.8	150	125 / 120	64 / 67	82	53	46	
Comoros	67	59.4 / 62.2	480	98 / 81	30 / 25	59	26	19	
Cyprus	8	76.0 / 80.5	47	97 / 97	93 / 94	10			
Djibouti	102	44.7 / 46.8	730	46 / 35	24 / 15	64			0.3 / 0.3
Equatorial Guinea	101	47.8 / 50.5	880	132 / 120	38 / 22	192			
Fiji	18	68.1 / 71.5	75	109 / 109	78 / 83	54			0.2 / <0.1
French Polynesia	9	70.7 / 75.8	20			45			
Guadaloupe	7	74.8 / 81.7	5			19			
Guam	10	72.4 / 77.0	12			70			
Guyana	51	60.1 / 66.3	170			67	37	36	2.2 / 2.8
Iceland	3	77.6 /81.9	0	101 / 101	104 / 111	19			0.2 / 0.2
Luxembourg	5	75.1 / 81.4	28	101 / 100	93 / 99	9			
Maldives	38	67.8 / 67.0	110	125 / 124	64 / 68	53			
Malta	7	75.9 / 80.7	21	106 / 106	91 / 89	12			
Martinique	7	75.8 / 82.3	4			30			
Micronesia (26)	21	70.3 / 74.0				53			
Netherlands Antilles	13	73.3 / 79.2	20	104 / 104	69 / 77	44			
New Caledonia	7	72.5 / 77.7	10			31			
Polynesia (27)	21	68.9 / 73.4				39			
Qatar	12	70.5 / 75.4	7	108 / 104	88 / 93	20	43	32	
Réunion	8	71.2 / 79.3	41			32	67	62	
Samoa	26	66.9 / 73.4	130	104 / 101	71 / 79	44			
Solomon Islands	21	67.9 / 70.7	130			52			
Suriname	26	68.5 / 73.7	110	127 / 125	62 / 86	42	42	41	2.2 / 1.1
Timor-Leste, Democratic Republic of	124	48.7 / 50.4	660			27			
Vanuatu	29	67.5 / 70.5	130	112 / 111	28 / 29	52			

Selected Indicators for Less Populous Countries/Territories

Demographic, Social and Economic Indicators	Total population (thousands) (2004)	Projected population (thousands) (2050)	% urban (2003)	Urban growth rate (2000-2005)	Population/ ha arable & perm. crop land	Total fertility rate (2000-2005)	% births with skilled attendants	GNI per capita PPP$ (2002)	Under-5 mortality M/F
Bahamas	317	395	89.5	1.5	0.9	2.29			27 / 21
Bahrain	739	1,270	90.0	2.3	1.2	2.66	98		20 / 16
Barbados	271	258	51.7	1.5	0.6	1.50	91		13 / 11
Belize	261	421	48.3	2.3	0.7	3.15	83	5,340	43 / 42
Brunei Darussalam	366	685	76.2	3.2	0.3	2.48	99		8 / 6
Cape Verde	473	812	55.9	3.5	2.4	3.30	89	4,720	45 / 26
Comoros	790	1,816	35.0	4.6	4.0	4.90	62	1,640	96 / 87
Cyprus	808	892	69.2	1.0	0.6	1.90	100	18,040	8 / 8
Djibouti	712	1,395	83.7	2.1		5.70		2,070	185 / 168
Equatorial Guinea	507	1,177	48.1	4.7	1.4	5.89	65	5,590	181 / 164
Fiji	847	969	51.7	2.5	1.1	2.88	100	5,310	21 / 23
French Polynesia	248	355	52.1	1.2		2.44			11 / 11
Guadaloupe	443	467	99.7	0.9	0.5	2.10			11 / 8
Guam	165	248	93.7	1.7		2.88			13 / 10
Guyana	767	507	37.6	1.4	0.3	2.31	86	3,780	81 / 60
Iceland	292	330	92.8	0.9	3.3	1.95		28,590	5 / 4
Luxembourg	459	716	91.9	1.6		1.73		51,060	7 / 7
Maldives	328	819	28.8	4.5	8.7	5.33	70		41 / 56
Malta	396	402	91.7	0.7	0.6	1.77			9 / 8
Martinique	395	413	95.7	0.8	0.7	1.90			9 / 8
Micronesia (26)	535	863	69.1	2.6		3.40			26 / 25
Netherlands Antilles	223	249	69.7	1.1	0.1	2.05			17 / 11
New Caledonia	233	382	61.2	2.2		2.45			9 / 10
Polynesia (27)	643	912	43.6	1.7		3.16			26 / 26
Qatar	619	874	92.0	1.7	0.3	3.22	98		17 / 13
Réunion	767	1,014	91.5	2.0	0.6	2.30			11 / 9
Samoa	180	254	22.3	1.3		4.12	100	5,350	34 / 29
Solomon Islands	491	1,071	16.5	4.5	4.4	4.42	85	1,520	31 / 30
Suriname	439	459	76.1	1.6	1.2	2.45	85		35 / 23
Timor-Leste, Democratic Republic of	820	1,433	7.6	4.8	7.2	3.85	24		186 / 179
Vanuatu	217	435	22.8	4.1		4.13	89	2,770	32 / 39

Notes for Indicators

The designations employed in this publication do not imply the expression of any opinion on the part of the United Nations Population Fund concerning the legal status of any country, territory or area or of its authorities, or concerning the delimitation of its frontiers or boundaries.

Data for small countries or areas, generally those with population of 200,000 or less in 1990, are not given in this table separately. They have been included in their regional population figures.

(*) More-developed regions comprise North America, Japan, Europe and Australia-New Zealand.

(+) Less-developed regions comprise all regions of Africa, Latin America and Caribbean, Asia (excluding Japan), and Melanesia, Micronesia and Polynesia.

(‡) Least-developed countries according to standard United Nations designation.

(1) Including British Indian Ocean Territory and Seychelles.

(2) Including Agalesa, Rodrigues and St. Brandon.

(3) Including Sao Tome and Principe.

(4) Formerly Zaire.

(5) Including Western Sahara.

(6) Including St. Helena, Ascension and Tristan da Cunha.

(7) Including Macau.

(8) On 1 July 1997, Hong Kong became a Special Administrative Region (SAR) of China.

(9) This entry is included in the more developed regions aggregate but not in the estimate for the geographical region.

(10) Turkey is included in Western Asia for geographical reasons. Other classifications include this country in Europe.

(11) Comprising Algeria, Bahrain, Comoros, Djibouti, Egypt, Iraq, Jordan, Kuwait, Lebanon, Libyan Arab Jamahiriya, Mauritania, Morocco, Occupied Palestinian Territory, Oman, Qatar, Saudi Arabia, Somalia, Sudan, Syria, Tunisia, United Arab Emirates and Yemen. Regional aggregation for demographic indicators provided by the UN Population Division. Aggregations for other indicators are weighted averages based on countries with available data.

(12) Including Channel Islands, Faeroe Islands and Isle of Man.

(13) Including Andorra, Gibraltar, Holy See and San Marino.

(14) Including Leichtenstein and Monaco.

(15) Including Anguilla, Antigua and Barbuda, Aruba, British Virgin Islands, Cayman Islands, Dominica, Grenada, Montserrat, Netherlands Antilles, Saint Kitts and Nevis, Saint Lucia, Saint Vincent and the Grenadines, Turks and Caicos Islands, and United States Virgin Islands.

(16) Including Falkland Islands (Malvinas) and French Guiana.

(17) Including Bermuda, Greenland, and St. Pierre and Miquelon.

(18) Including Christmas Island, Cocos (Keeling) Islands and Norfolk Island.

(19) Including New Caledonia and Vanuatu.

(20) The successor States of the former USSR are grouped under existing regions. Eastern Europe includes Belarus, Republic of Moldova, Russian Federation and Ukraine. Western Asia includes Armenia, Azerbaijan and Georgia. South Central Asia includes Kazakhstan, Kyrgyzstan, Tajikistan, Turkmenistan and Uzbekistan. Regional total, excluding subregion reported separately below.

(21) Regional total, excluding subregion reported separately below.

(22) These subregions are included in the UNFPA Arab States and Europe region.

(23) Estimates based on previous years' reports. Updated data are expected.

(24) Total for Eastern Europe includes some South European Balkan States and Northern European Baltic States.

(25) More recent reports suggest this figure might have been higher. Future publications will reflect the evaluation of this information.

(26) Comprising Federated States of Micronesia, Guam, Kiribati, Marshall Islands, Nauru, Northern Mariana Islands, and Pacific Islands (Palau).

(27) Comprising American Samoa, Cook Islands, Johnston Island, Pitcairn, Samoa, Tokelau, Tonga, Midway Islands, Tuvalu, and Wallis and Futuna Islands.

Technical Notes

The statistical tables in this year's *State of World Population* report once again give special attention to indicators that can help track progress in meeting the quantitative and qualitative goals of the International Conference on Population and Development (ICPD) and the Millennium Development Goals (MDGs) in the areas of mortality reduction, access to education, access to reproductive health services including family planning, and HIV/AIDS prevalence among young people. Several changes have been made in other indicators, as noted below. Future reports will include different process measures when these become available, as ICPD and MDG follow-up efforts lead to improved monitoring systems. Improved monitoring of the financial contributions of governments, non-governmental organizations and the private sector should also allow better future reporting of expenditures and resource mobilization for ICPD/MDG implementation efforts. The sources for the indicators and their rationale for selection follow, by category.

Monitoring ICPD Goals

INDICATORS OF MORTALITY

Infant mortality, male and female life expectancy at birth. Source: United Nations Population Division. 2003. *World Population Prospects: The 2002 Revision.* New York: United Nations. Spreadsheets provided by United Nations Population Division. These indicators are measures of mortality levels, respectively, in the first year of life (which is most sensitive to development levels) and over the entire lifespan. Estimates are for the 2000-2005 period.

Maternal mortality ratio. Source: WHO, UNICEF, and UNFPA. 2003. *Maternal Mortality in 2000: Estimates Developed by WHO, UNICEF, and UNFPA.* Geneva: WHO. This indicator presents the number of deaths to women per 100,000 live births which result from conditions related to pregnancy, delivery and related complications. Precision is difficult, though relative magnitudes are informative. Estimates below 50 are not rounded; those 50-100 are rounded to the nearest 5; 100-1,000, to the nearest 10; and above 1,000, to the nearest 100. Several of the estimates differ from official government figures. The estimates are based on reported figures wherever possible, using approaches to improve the comparability of information from different sources. See the source for details on the origin of particular national estimates. Estimates and methodologies are

reviewed regularly by WHO, UNICEF, UNFPA, academic institutions and other agencies and are revised where necessary, as part of the ongoing process of improving maternal mortality data. Because of changes in methods, prior estimates for 1995 levels may not be strictly comparable with these estimates.

INDICATORS OF EDUCATION

Male and female gross primary enrolment ratios, male and female gross secondary enrolment ratios. Source: Spreadsheet provided by the UNESCO Institute for Statistics, March 2004. Data for countries from the OECD database are provisional (see details at: www1.oecd.org/els/education/ei/eag/wei.htm, last accessed 1 June 2004). Population data is based on: United Nations Population Division, *World Population Prospects: The 2002 Revision.* Gross enrolment ratios indicate the number of students enrolled in a level in the education system per 100 individuals in the appropriate age group. They do not correct for individuals who are older than the level-appropriate age due to late starts, interrupted schooling or grade repetition. Data are for 2001/2002 year, or for 2000/2001 if later date not available.

Male and female adult illiteracy. Source: See gross enrolment ratios above for source; data adjusted to illiteracy, from literacy. Illiteracy definitions are subject to variation in different countries; three widely accepted definitions are in use. Insofar as possible, data refer to the proportion who cannot, with understanding, both read and write a short simple statement on everyday life. Adult illiteracy (rates for persons above 15 years of age) reflects both recent levels of educational enrolment and past educational attainment. The above education indicators have been updated using the UN Population Division estimates from *World Population Prospects: The 2002 Revision.* New York: United Nations. Data are for the most recent year estimates available for the 2000-2004 period.

Proportion reaching grade 5 of primary education. Source: See gross enrolment ratios above for source. Data are most recent within the school years beginning in 1999, 2000, or 2001. Twenty-three countries reported data to grade 4 (see original source).

INDICATORS OF REPRODUCTIVE HEALTH

Births per 1,000 women aged 15-19. Source: Spreadsheets provided by the United Nations Population Division. This is an indicator of the burden of fertility on young women.

Since it is an annual level summed over all women in the age cohort, it does not reflect fully the level of fertility for women during their youth. Since it indicates the annual average number of births per woman per year, one could multiply it by five to approximate the number of births to 1,000 young women during their late teen years. The measure does not indicate the full dimensions of teen pregnancy as only live births are included in the numerator. Stillbirths and spontaneous or induced abortions are not reflected.

Contraceptive prevalence. Source: Spreadsheet, "Percent Currently Using Contraception among Married or In-union Women of Reproductive Age", provided by United Nations Population Division using "World Contraceptive Use 2003: Database Maintained by the Population Division of the United Nations Secretariat." These data are derived from sample survey reports and estimate the proportion of married women (including women in consensual unions) currently using, respectively, any method or modern methods of contraception. Modern or clinic and supply methods include male and female sterilization, IUD, the pill, injectables, hormonal implants, condoms and female barrier methods. These numbers are roughly but not completely comparable across countries due to variation in the timing of the surveys and in the details of the questions. Unlike in past years, all country and regional data refer to women aged 15-49. The most recent survey data available are cited, ranging from 1980-2002.

HIV prevalence rate, M/F, 15-49. Source: Data provided by UNAIDS. UNAIDS 2004. Geneva: UNAIDS. These data derive from surveillance system reports and model estimates. Data provided for men and women aged 15-49 are point estimates for each country. The reference year is 2003. Male-female differences reflect physiological and social vulnerability to the illness and are affected by age differences between sexual partners.

Demographic, Social and Economic Indicators

Total population 2003, projected population 2050, average annual population growth rate for 2000-2005. Source: Spreadsheets provided by United Nations Population Division. These indicators present the size, projected future size and current period annual growth of national populations.

Per cent urban, urban growth rates. Source: United Nations Population Division. 2004. *World Urbanization Prospects: The 2003 Revision.* New York: United Nations, available from CD-ROM (POP/DP/WUP/Rev.2003), and United Nations Population Division. 2004. *World*

Urbanization Prospects: The 2003 Revision: Data Tables and Highlights (ESA/P/WP.190). New York: United Nations. These indicators reflect the proportion of the national population living in urban areas and the growth rate in urban areas projected.

Agricultural population per hectare of arable and permanent crop land. Source: Data provided by Food and Agriculture Organization, Statistics Division, using agricultural population data based on the total populations from United Nations Population Division. 2003. *World Population Prospects: The 2002 Revision.* New York: United Nations. This indicator relates the size of the agricultural population to the land suitable for agricultural production. It is responsive to changes in both the structure of national economies (proportions of the workforce in agriculture) and in technologies for land development. High values can be related to stress on land productivity and to fragmentation of land holdings. However, the measure is also sensitive to differing development levels and land use policies. Data refer to the year 2001.

Total fertility rate (period: 2000-2005). Source: United Nations Population Division. 2003. *World Population Prospects: The 2002 Revision.* New York: United Nations. The measure indicates the number of children a woman would have during her reproductive years if she bore children at the rate estimated for different age groups in the specified time period. Countries may reach the projected level at different points within the period.

Births with skilled attendants. Source: Spreadsheet provided by UNICEF, with data from *State of World's Children 2004* and February 2004 MDG Monitoring updates. Data for more developed countries are not available. This indicator is based on national reports of the proportion of births attended by "skilled health personnel or skilled attendant: doctors (specialist or non-specialist) and/or persons with midwifery skills who can diagnose and manage obstetrical complications as well as normal deliveries". Data for more developed countries reflect their higher levels of skilled delivery attendance. Because of assumptions of full coverage, data (and coverage) deficits of marginalized populations and the impacts of chance and transport delays may not be fully reflected in official statistics. Data estimates are the most recent available after 1994.

Gross national income per capita. Source: Most recent (2001 or 2002) figures from: The World Bank. *World Development Indicators Online.* Web site: http://devdata.worldbank.org/dataonline/ (by subscription). This indicator (formerly referred to as gross national product [GNP] per capita) measures the total output of goods and services for final use produced by residents and non-residents, regardless of allocation to domestic and foreign claims, in relation to the size of the population. As such, it is an indicator of the economic productivity of a nation. It differs from gross domestic product (GDP) by further adjusting for income received from abroad for labour and capital by residents, for similar payments to non-residents, and by incorporating various technical adjustments including those related to exchange rate changes over time. This measure also takes into account the differing purchasing power of currencies by including purchasing power parity (PPP) adjustments of "real GNP". Some PPP figures are based on regression models; others are extrapolated from the latest International Comparison Programme benchmark estimates; see original source for details.

Central government expenditures on education and health. Source: Most recent data point in last 6 years from: The World Bank. *World Development Indicators Online.* Web site: http://devdata.worldbank.org/dataonline/ (by subscription). These indicators reflect the priority afforded to education and health sectors by a country through the government expenditures dedicated to them. They are not sensitive to differences in allocations within sectors, e.g., primary education or health services in relation to other levels, which vary considerably. Direct comparability is complicated by the different administrative and budgetary responsibilities allocated to central governments in relation to local governments, and to the varying roles of the private and public sectors. Reported estimates are presented as shares of GDP per capita (for education) or total GDP (for health). Great caution is also advised about cross-country comparisons because of varying costs of inputs in different settings and sectors.

External assistance for population. Source: UNFPA. 2003. *Financial Resource Flows for Population Activities in 2001.* New York: UNFPA. This figure provides the amount of external assistance expended in 2001 for population activities in each country. External funds are disbursed through multilateral and bilateral assistance agencies and by non-governmental organizations. Donor countries are indicated by their contributions being placed in parentheses. Regional totals include both country-level projects and regional activities (not otherwise reported in the table). Data for 2002 will be available post-publication.

Under-5 mortality. Source: United Nations Population Division, special tabulation based on United Nations. 2003. *World Population Prospects: The 2002 Revision.* New York: United Nations. This indicator relates to the incidence of mortality to infants and young children. It reflects, therefore, the impact of diseases and other causes of death on infants, toddlers and young children. More standard demographic measures are infant mortality and mortality rates for 1 to 4 years of age, which reflect differing causes of and frequency of mortality in these ages. The measure is more sensitive than infant mortality to the burden of childhood diseases, including those preventable by improved nutrition and by immunization programmes. Under-5 mortality is here expressed as deaths to children under 5 per 1,000 live births in a given year. The estimate refers to the period 2000-2005.

Per capita energy consumption. Source: The World Bank. *World Development Indicators Online.* Web site: http://devdata.worldbank.org/dataonline/ (by subscription). This indicator reflects annual consumption of commercial primary energy (coal, lignite, petroleum, natural gas and hydro, nuclear and geothermal electricity) in kilograms of oil equivalent per capita. It reflects the level of industrial development, the structure of the economy and patterns of consumption. Changes over time can reflect changes in the level and balance of various economic activities and changes in the efficiency of energy use (including decreases or increases in wasteful consumption). Data estimates are for 2001.

Access to safe water. Source: UNICEF. 2003. *The State of the World's Children 2004: Girls, Education and Development.* New York: UNICEF: Table 3: Health. This indicator reports the percentage of the population with access to an *improved source* of drinking water providing *adequate amount of safe water* located within a *convenient distance* from the user's dwelling. The italicized words use country-level definitions. The indicator is related to exposure to health risks, including those resulting from improper sanitation. Data are estimates for the year 2000.

Editorial Team

The State of World Population 2004

Editor: William A. Ryan
Senior Researcher/Policy Adviser: Stan Bernstein
Editorial Assistant: Phyllis Brachman
Contributors: David Del Vecchio, Lucille Pilling de Lucena, Patrick Friel, Margaret E. Greene, Karen Hardee, Marianne Haslegrave, Erin Hasselberg, Don Hinrichsen, Mia MacDonald, Alex Marshall, Kourtoum Nacro, Danielle Nierenberg, Rabbi Royan, Gita Sen, Michael Vlassoff
Intern: Katherine McCarthy

Prepress/Production: Prographics, Inc., Annapolis, Maryland, USA

Photo captions and credits

Front cover
© Ron Giling/Still Pictures
Mothers and children in Ghana.

Chapter 1
© Dominic Sansoni/Panos Pictures
Rural family in India.

Chapter 2
© Mark Edwards/Still Pictures
Migrant family in the Philippines.

Chapter 3
© Mark Edwards/Still Pictures
Workers in Wollo, Ethiopia, divert a stream to irrigate land as part of a Concern Food for Work project.

Chapter 4
© Hans Blossey/Still Pictures
People and cars fill Nathan Road in Hong Kong SAR, China.

Chapter 5
© Roger LeMoyne/UNICEF
Teachers attend a Ministry of Education training session, Kabul, Afghanistan.

Chapter 6
© Shehzad Noorani/Still Pictures
A community health worker gives family planning education to a rural woman in Bangladesh.

Chapter 7
© Shehzad Noorani/Still Pictures
Prenatal exam in India.

Chapter 8
© Mark Edwards/Still Pictures
Outreach worker distributes condoms to prostitutes in Bangkok, Thailand.

Chapter 9
© Jorgen Schytte/Still Pictures
Young mother and infant in Guatemala.

Chapter 10
© Mark Edwards/Still Pictures
Father and child in India.

Chapter 11
© Shehzad Noorani/Still Pictures
Mother and daughter in Bangladesh.

Page 20
© Mark Edwards/Still Pictures
Migrant family works to turn patches of rainforest into land for subsistence farming, Java, Indonesia.

Page 59
© Giacomo Pirozzi/Panos Pictures
Nurse examines a pregnant woman at clinic in Cape Verde.

Page 66
© Marie Dorigny
AIDS patient receives care at Bukoba Hospital, Tanzania.

United Nations Population Fund
220 East 42nd Street, 23rd Fl.
New York, NY 10017 U.S.A.
www.unfpa.org